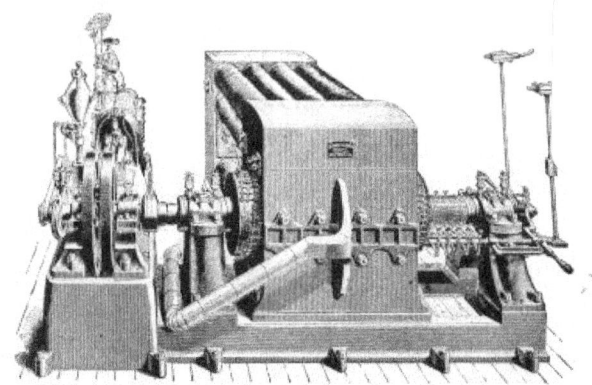

FIRST DIRECT-CONNECTED ELECTRIC GENERATOR UNIT OF LARGE CAPACITY EVER CONSTRUCTED UP TO THE TIME IT WAS MADE BY THOMAS A. EDISON IN JUNE, 1881. CAPACITY, 1200 INCANDESCENT LAMPS OF 16 CANDLE-POWER EACH

A-B-C
OF
ELECTRICITY

BY
WILLIAM H. MEADOWCROFT

HARPER & BROTHERS PUBLISHERS

NEW YORK & LONDON

A-B-C of Electricity

From the Laboratory of Thomas A. Edison

Orange, N. J.

Mr. W. H. Meadowcroft,

New York City.

DEAR SIR:

I have read the MS. of your "A-B-C of Electricity," and find that the statements you have made therein are correct. Your treatment of the subject, and arrangement of the matter, have impressed me favorably.

Yours truly,

THOS. A. EDISON

CONTENTS

[vii]
[viii]

INTRODUCTION TO NEW EDITION

The favor with which this book has been received has brought about the preparation of this new edition. The present volume has been enlarged by the addition of certain new material and it has been entirely reset. Some new illustrations have been made, and in its new dress the book, it is hoped, will be found to afford an even larger measure of usefulness. The principles of the science remain the same, but the author is glad of the opportunity to note certain developments in their application.

W. H. M.

Edison Laboratory, *April, 1915.*

PREFACE

While there is no lack of most excellent text-books for the study of those branches of Electricity which are above the elementary stage, there is a decided need of text-books which shall explain, in simple language, to young people of, say, fourteen years and upward, a general outline of the science, as well as the ground-work of those electrical inventions which are to-day of such vast commercial importance.

There is also a need for such a book among a large part of the adult population, for the reason that there have been great and radical changes in this science since the time they completed their studies, and they have not the time to follow up the subject in the advanced books.

As instances of those changes just spoken of, the electric light, telephone, and storage batteries may be mentioned, which have been developed during the last ten or twelve years, with the result of adding very many features that were entirely new to electricians.

With these ideas in view I have prepared this little volume. It is not intended, in the slightest degree, to be put forward as a scientific work, but it will probably give to many the information they desire without requiring too great a research into books which treat more extensively and deeply of this subject.

W. H. M.

A-B-C OF ELECTRICITY

A-B-C OF ELECTRICITY

I

We now obtain so many of our comforts and conveniences by the use of electricity that all young people ought to learn something of this wonderful force, in order to understand some of the principles which are brought into practice.

You all know that we have the telegraph, the telephone, the electric light, electric motors on street-cars, electric bells, etc., besides many other conveniences which the use of electricity gives us.

Every one knows that, by the laws of multiplication, twice two makes four, and that twice two can never make anything but four. Well, these useful inventions have been made by applying the *laws of electricity* in certain ways, just as well known, so as to enable us to send in a few moments a message to our absent friends at any distance, to speak with them at a great distance, to light our houses and streets with electric light, and to do many other useful things with quickness and ease.

But you must remember that we do not know what electricity itself really is. We only know how to produce it by certain methods, and we also know what we can do with it when we have obtained it.

In this little book we will try to explain the various ways by which electricity is obtained, and how it is applied to produce the useful results that we see around us.

We will try and make this explanation such that it will encourage many of you to study this very important and interesting subject more deeply.

In the advanced books on electricity there are many technical terms which are somewhat difficult to understand, but in this book it will only be necessary to use a few of the more simple ones, which it will be well for you to learn and understand before going further.

II

DEFINITIONS

The three measurements most frequently used in electricity are

The Volt,
The Ampère,
The Ohm.

We will explain these in their order.

Fig. 1

The Volt.—This term may be better understood by making a comparison with something you all know of. Suppose we have a tank containing one hundred gallons of water, and we want to discharge it through a half-inch pipe at the bottom of the tank. Suppose, further, that we wanted to make the water spout upward, and for this purpose the pipe was bent upward as in Fig. 1.

If you opened the tap the water would spout out and upward as in Fig. 1.

Fig. 2

The cause of its spouting upward would be the *weight* or *pressure* of the water in the tank. This pressure is reckoned as so many *pounds* to the square inch of water.

Now, if the tank were placed on the roof of the house and the pipe brought to the ground as shown in Fig. 2, the water would spout up very much higher, because there would be *many more pounds* of pressure on account of the height of the pipe.

So, you see, the force or pressure of water is measured in pounds, and, therefore, a pound is the unit of pressure, or force, of water. Now, in electricity the unit of pressure, or force, is called a volt.

This word "volt" does not mean any weight, as the word "pound" weight does. You all know that if you have a pound of water you must have something to hold it, because it has weight, and, consequently, occupies some space. But *electricity itself has no weight* and therefore cannot occupy any space.

When we desire to carry water into a house or other building we do so by means of hollow pipes, which are usually made of iron. This is the way that water is brought into houses in cities and towns, so that it may be drawn and used in any part of a dwelling. Now, the principal supply usually comes from a reservoir which is placed up on high ground so as to give the necessary pounds of pressure to force the water up to the upper part of the houses. If some arrangement of this kind were not made we could get no water in our bedrooms, because, as you know, water will not rise above its own level unless by force.

The water cannot escape as long as there are no holes or leaks in the iron pipes, but if there should be the slightest crevice in them the water will run out.

In electricity we find similar effects.

The electricity is carried into houses by means of wires which are covered, or *insulated*, with various substances, such, for instance, as rubber. Just as the iron of the pipes prevents the water from escaping, the insulation of the wire prevents the escape of the electricity.

Now, if we were to cause the pounds of pressure of water, in pipes of ordinary thickness, to be very greatly increased, the pipes could not stand the strain and would burst and the water escape. So it is with electricity. If there were too many volts of pressure the insulation would not be sufficient to hold it and the electricity would escape through the covering, or insulation, of the wire.

It is a simple and easy matter to stop the flow of water from an ordinary faucet by placing your finger over the opening. As the water cannot then flow, your finger is what we will call a non-conductor and the water will be retained in the pipe.

We have just the same effects in electricity. If we place some substance which is practically a non-conductor, or insulator, such as rubber, around an electric wire, or in the path of an electric current, the electricity, acted upon by the volts of pressure, cannot escape, because the insulation keeps it from doing so, just as the iron of the pipe keeps the water from escaping. Thus, you see, the volt does not itself represent electricity, but only the pressure which forces it through the wire.

There are other words and expressions in electricity which are sometimes used in connection with the word "volt." These words are "pressure" and "intensity." We might say, for instance, that a certain dynamo machine had an electromotive force of 110 volts; or that the intensity of a cell of a battery was 2 volts, etc.

We might mention, as another analogy, the pressure of steam in a boiler, which is measured or calculated in pounds, just as the pressure of water is measured. So, we might say that 100 pounds steam pressure used through the medium of a steam-engine to drive a dynamo could thus be changed to electricity at 100 volts pressure.

The Ampère.—Now, in comparing the pounds pressure of water with the volts of pressure of electricity we used as an illustration a tank of water containing 100 gallons, and we saw that this water had a downward force or pressure in pounds. Let us now see what this pressure was acting upon.

It was forcing the quantity of water to spout upward through the end of the pipe. Now, as the quantity of water was 100 gallons, it could not all be forced at once out of the end of the pipe. The pounds pressure of water acting on the 100 gallons would force it out at a *certain rate*, which, let us say, would be one gallon per minute.

This would be the *rate of the flow* of water out of the tank.

Thus, you see, we find a second measurement to be considered in discharging the water-tank. The first was the force, or pounds of pressure, and the second the *rate* at which the quantity of water was being discharged per minute by that pressure.

This second measurement teaches us that a *certain quantity* will pass out of the pipe in a *certain time* if the pressure is steady, such quantity depending, of course, on the size or friction resistance of the pipe.

In electricity the volts of pressure act so as to force the quantity of current to *flow through the wires at a certain rate* per second, and the rate at which it flows is measured in ampères. For instance, let us suppose that an electric lamp required a pressure of 100 volts and a current of one ampère to light it up, we should have to supply a current of electricity flowing at the rate of one ampère, acted upon by an electromotive force of 100 volts.

You will see, therefore, that while the volt does not represent any electricity, but only its pressure, the ampère represents the *rate of flow* of the current itself.

You should remember that there are several words sometimes used in connection with the word "ampère"—for instance, we might say that a lamp required a "current" of one ampère or that a dynamo would give a "quantity" of 20 ampères.

The Ohm.—You have learned that the *pressure* would discharge the *quantity* of water at a certain rate through the pipe. Now, suppose we were to fix *two* discharge-pipes to the tank, the water would run away very much quicker, would it not? If we try to find a reason for this, we shall see that a pipe can only, at a given pressure, admit so much water through it at a time.

Therefore, you see, this pipe would present a certain amount of *resistance* to the passage of the total quantity of water, and would only allow a limited quantity at once to go through. But, if we were to attach two or more pipes to the tank, or one large pipe, we should make it easier for the water to flow, and, therefore, the total amount of resistance to the passage of the water would be very much less, and the tank would quickly be emptied.

Now, as you already know, water has substance and weight and therefore occupies some space, but electricity has neither substance nor weight, and therefore cannot occupy any space; consequently, to carry electricity from one place to another we do not need to use a pipe, which is hollow, but we use a solid wire.

These solid wires have a certain amount of *resistance* to the passage of the electricity, just as the water-pipe has to the water, and (as it is in the case of the water) the effect of the resistance to the passage of electricity is greater if you pass a larger quantity through than a smaller quantity.

If you wanted to carry a quantity of electricity to a certain distance, and for that purpose used a wire, there would be a certain amount of resistance in that wire to the passage of the current through it; but if you used two or more wires of the same size, or one large wire, the resistance would be very much less and the current would flow more easily.

Suppose that, instead of emptying the water-tank from the roof through the pipe, we had just turned the tank over and let the water all pour out at once down to the ground. That would dispose of the water very quickly and by a short way, would it not? That is very easy to be seen, because there would be *no resistance* to its passage to the ground.

Well, suppose we had an electric battery giving a certain quantity of current, say five ampères, and we should take a large wire that would offer no resistance to that quantity and put it from one side of the battery to the other, a large current would flow at once and tend to exhaust the battery. This is called a *short circuit* because there is little or no resistance, and it provides the current with an easy path to escape. Remember this, that *electricity always takes the easiest path*. It will take as many paths as are offered, but the largest quantity will always take the easiest.

As the subject of resistance is one of the most important in electricity, we will give you one more example, because if you can obtain a good understanding of this principle it will help you to comprehend the whole subject more easily in your future studies.

We started by comparison with a tank holding 100 gallons of water, discharging through a half-inch pipe, and showed you that the pounds of pressure would force the quantity of gallons through the pipe. When the tap was first opened the water would spout up very high, but as the water in the tank became lower the pressure would be less, and, consequently, the water would not spout so high.

So, if it were desired to keep the water spouting up to the height it started with, we should have to keep the tank full, so as to have the same pounds of pressure all the time. But, if we wanted the water to spout still higher we should have to use other means, such as a force-pump, to obtain a greater pressure.

Now, if we should use too many pounds pressure it would force the quantity of water more rapidly through the pipe and would cause the water to become heated because of the resistance of the pipe to the passage of that quantity acted upon by so great a pressure.

This is just the same in electricity, except that the wire itself would become heated, some of the electricity being turned into heat and lost. If a wire were too small for the volts pressure and ampères of current of electricity the resistance of such wire would be overcome, and it would become red-hot and perhaps melt. Electricians are therefore very careful to calculate the resistance of the wires they use before putting them up, especially when they are for electric lighting, in order to make allowances for the ampères of current to flow through them, so that but little of the electricity will be turned into heat and thus rendered useless for their purpose.

The unit of resistance is called the *ohm* (pronounced like "home" without the "h").

All wires have a certain resistance per foot, according to the nature of the metal used and the size of the wire—that is to say, the finer the wire the greater number of ohms resistance it has to the foot.

Water and electricity flow under very similar conditions—that is to say, each of them must have a channel, or conductor, and each of them requires pressure to force it onward. Water, however, being a tangible substance, requires a hollow conductor; while electricity, being intangible, will flow through a solid conductor. The iron of the water-pipe and the insulation of the electric wire serve the same purpose—namely, that of serving to prevent escape by reason of the pressure exerted.

There is another term which should be mentioned in connection with resistance, as they are closely related, and that is *opposition*. There is no general electrical term of this name, but, as it will be most easily understood from the meaning of the word itself, we have used it.

Let us give an example of what opposition would mean if applied to water. Probably every one knows that a water-wheel is a wheel having large blades, or "paddles," around its circumference.

When the water, in trying to force its passage, rushes against one of these paddles it meets with its opposition, but overcomes it by pushing the paddle away. This brings around more opposition in the shape of another paddle, which the water also pushes away. And so this goes on, the water overcoming this opposition and turning the wheel around, by which means we can get water to do useful work for us.

You must remember, however, that it is only by putting opposition in the path of a pressure and quantity of water that we can get this work.

The same principle holds good in electricity. We make electricity in different ways, and in order to obtain useful work we put in its path the instruments, lamps, or machines which offer the proper amount of resistance, or opposition, to its passage, and thus obtain from this wonderful agent the work we desire to have done.

You have learned that three important measurements in electricity are as follows:

The *volt* is the practical unit of measurement of *pressure*;

The *ampère* is the practical unit of measurement of the *rate of flow*; and

The *ohm* is the practical unit of measurement of *resistance*.

III

MAGNETISM

Now we will try to explain to you something about magnets and magnetism. There are very few boys who have not seen and played with the ordinary magnets, shaped like a horseshoe, which are sold in all toy-stores as well as by those who sell electrical goods.

Well, you know that these magnets will attract and hold fast anything that is made of iron or steel, but they have no effect on brass, copper, zinc, gold, or silver, yet there is nothing that you can see which should cause any such effect. You will notice, then, that magnetism is like electricity; we cannot see it, but we can tell that it exists, because it produces certain effects. And here is another curious thing—magnetism produces electricity, and electricity produces magnetism. This seems to be a very convenient sort of a family affair, and it is owing to this close relation that we are able to obtain so many wonderful things by the use of electricity.

We shall now show you how electricity produces magnetism, and, when we come to the subject of electric lighting we will explain how magnetism produces electricity.

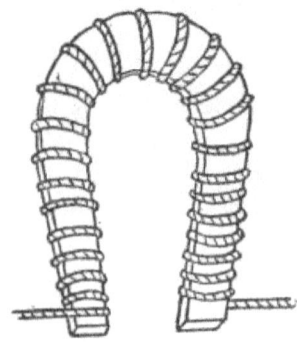

Fig. 3

The easiest way to show how electricity makes magnetism is to find out how magnets are made. Suppose we wanted to make a horseshoe magnet, just mentioned above; we would take a piece of *steel* and wind around it some fine copper wire, commencing on one leg of the horseshoe and winding around until we came to the end of the other leg. Then we should have two ends of wire left, as shown in the sketch. (Fig. 3.)

We connect these two ends with an electric battery, giving, say, two volts, and then the ampères of current of electricity will travel through the wire, and in doing so has such an influence on the steel that it is converted into a magnet, such as you have played with. The current is "broken"—that is to say, it is shut off several times in making a magnet of this kind, and then the wire is taken away from the battery and is unwound from the steel horseshoe, leaving it free from wire, just as you have seen it. This horseshoe is now a *permanent magnet*—that is, it will *always* attract and hold pieces of iron and steel.

Now, if you were to do the same thing with a horseshoe made of soft iron instead of steel it would not be a magnet after you stopped the current of electricity from going through the wires, although the piece of *iron* would be a stronger magnet while the electricity was going through the wire around it.

The steel magnet is called a permanent magnet, and its ends, or "poles," are named North and South. There is usually a loose piece of steel or iron, called an "armature," put across the ends, which has the peculiar property of keeping the magnetism from becoming weaker, and thereby retaining the strength of the magnet. The strongest part of the magnet is at the poles, while, at the point marked + (which is called the neutral point) there is scarcely any magnetism.

It will be well to remember the object of the *armature* as we shall meet it again in describing dynamo machines.

The magnets made of iron are called electromagnets because they exhibit magnetism only when the ampères of current of electricity are flowing around them. They also have two poles, north and south, as have permanent magnets. Electromagnets are used in nearly all electrical instruments, not only because they are stronger than permanent magnets, but because they can be made to act instantly by passing a current of electricity through them at the most convenient moment, as you will see when we explain some of the electrical instruments which are used to produce certain effects. (Fig. 4.)

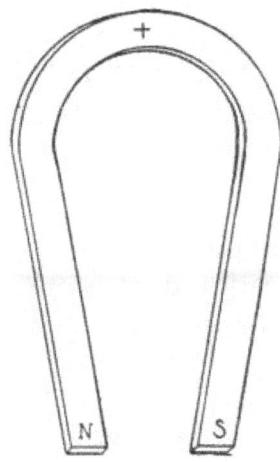

Fig. 4

Of course there are a great many different shapes in which magnets are made. The simplest is the *bar magnet*, which is simply a flat or round piece of iron or steel. Suppose you made a magnet of a flat piece of steel and put on top of it a sheet of paper, and then threw on the paper some iron filings, you would see them arrange themselves as is shown in the following sketch. (Fig. 5.)

The filings would always arrange themselves in this shape, no matter how large or small the magnets were. And, if you were to cut it into two or half a dozen pieces, each piece would have the same effect. This shows you that each piece would itself become a magnet and would have its poles exactly as the large one had.

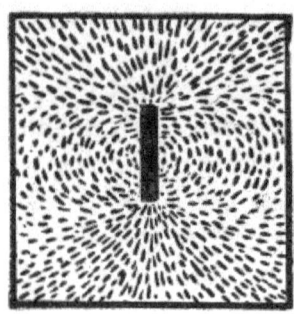

Fig. 5

Now, we have another curious thing to tell you about magnets. If you present the north pole of a magnet to the south pole of another magnet, they will attract and hold fast to each other, but if you present a south pole to another south pole, or a north pole to a north pole, they will repel each other, and there will be no attraction. You can perform some interesting experiments by reason of this fact. We will give you one of them.

Take, say, a dozen needles and draw them several times in the same direction across the ends of a magnet so that they become magnetized. Now stick each needle half-way through a piece of cork, and put the corks, with the needles sticking through them, into a bowl of water. Then take a bar magnet and bring it gradually toward the middle of the bowl and you will see the corks advance or back away from the magnet. If the ends of the needles sticking up out of the water are south poles and the end of the magnet you present is a north pole, the needles will come to the center; but will go to the side of the bowl if you present the south pole. You can vary this pretty experiment by turning up the other ends of part of the needles.

You will remember that when we explained what "resistance" meant, we told you that electricity would always take the easiest path, and while part of it will flow in a small wire, the largest portion will take an easier path if it can get to something larger that is a metallic substance. Electricity will only flow easily through anything that is made of metal. You will also remember that you learned that when electricity took a short cut to get away from its proper path it was called a *short circuit*.

All this must be taken into consideration when magnets are being made. In the first place, the wire we wind around steel or iron to make magnets must always be covered with an insulator of electricity. Magnet wire is usually covered with cotton or silk. If it were left bare, each turn of the wire would touch the next turn, and so we should make such an easy path for the electricity that it would all go back to the battery by a short circuit, and then we would get no magnetic effect in the steel or iron. *The only way we can get electricity to do useful work for us is to put some resistance or opposition in its way.* So you see that if we make it travel through the wire around the iron or steel, there is just enough resistance or opposition in its way to give it work to get through the wire, and this work produces the peculiar effect of making the iron or steel magnetic.

The covering on the wire, as you will remember, is called "insulation."

IV

THE TELEGRAPH

Every one knows how very convenient the telegraph is, but there are not many who think how wonderful it is that we can send a message in a few seconds of time to a distant place, even though it were thousands of miles away. And yet, though the present system of telegraphing is a wonderful one, the method of sending a telegram is simple enough. The apparatus that is used in sending a telegram is as follows:

The Battery.
The Wire.
The Telegraph Key.
The Sounder.

The different kinds of electric batteries will be mentioned afterward, so we will not stop now to describe them, but simply state that a battery is used to produce the necessary electricity. As you all know what wire is, there is no necessity of describing it further.

The telegraph key is shown in the sketch below. (Fig. 6.)

Fig. 6

This instrument is usually made of brass, except that upon the handle there is the little knob which is of hard rubber. The handle, or lever, moves down when this knob is pressed, and a little spring beneath pushes it up again when let go. You will see a second smaller knob, the use of which we will explain later.

The sounder is shown on the following page. (Fig. 7.)

The part consisting of the two black pillars is an electromagnet, and across the top of these pillars is a piece of iron called the "armature," which is held up by a spring.

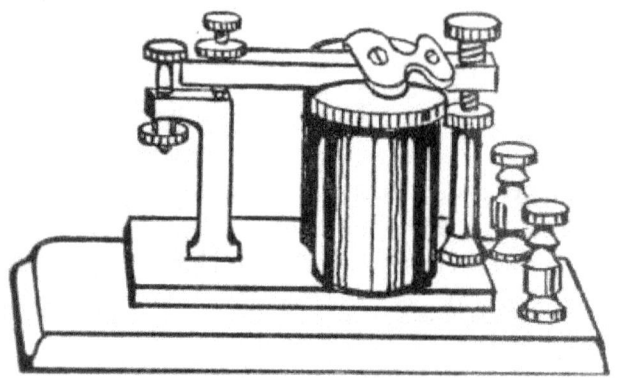

Fig. 7

Now let us see how the battery and wire are placed in connection with these instruments. You have seen that we usually have two wires for the electricity to travel in, one wire for it to leave the battery, and the other to return on. But you will easily see that if two wires had to be used in telegraphing it would be a very expensive matter, especially when they had to be carried thousands of miles. So, instead of using a second wire, we use the earth to carry back the electricity to the battery, because the earth is a better conductor even than wire. Although a quantity of ground equal in size to the wire would offer thousands of times greater resistance than the wire, yet, owing to the great body of our earth, its total resistance is even less than any telegraph wire used.

When two electric wires are run from a battery and connected together through some instrument, this is called a "circuit," because the electricity has a path in which it can travel back to the battery. This would be a "metallic" circuit; *but when one wire only* is used, and the other side of the battery is connected with the earth, it is called a "ground" or "earth" circuit, because the electricity returns through the earth.

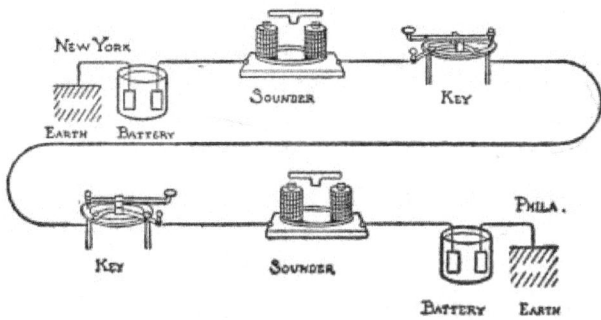

Fig. 8

If you look at this sketch (Fig. 8) you will see how the telegraph instruments are connected and will then be able to understand how a message can be sent.

Here we have two sets of telegraph apparatus, one of which, let us say, is in New York and the other in Philadelphia.

You will see that one wire from the battery is connected with the earth, and the other wire with the sounder. Another wire goes from the sounder to one leg of the key so as to make the brass base of the key part of the circuit. The other leg of the key is "insulated" from the brass base by being separated therefrom with some substance which will not carry electricity, such, for instance, as hard rubber.

We will suppose that there is already a wire strung up on poles between New York and Philadelphia, and that the key, sounder, and battery in the latter city are connected in the same way as those in New York.

Now, to enable us to send a message from one city to the other we must connect the ends of the wires to the instruments in each city; so we connect one end to the insulated leg of the key in New York, and the other end to the insulated leg of the key in Philadelphia.

Everything is now completed, and, as soon as we find out what is the use of that part of the key that has a little round, black handle, we shall be ready to start. This is called the "switch."

If you will look once more at the picture of the key you will see under the long handle (or lever) a little point which the lever will touch when it is pressed down. Now this little point is part of that insulated leg, and, therefore, this point is also insulated from the base. If a current of electricity were sent along the wire it could not get any farther than this point unless we put in some arrangement to complete the path, or circuit, for it to travel in. We therefore put in the switch.

One end of the switch (which is made of brass with a rubber handle) is fastened on the base of the key, so that it may be moved to the right or left. The other end, when the switch is moved to the left (or "closed"), touches a piece of brass fastened to the little point we have mentioned, and so makes a free path for the electricity to go through the base of the key and through the wire to the sounder, and from there to the battery, and so back to the earth. This switch must be opened before the sounder near it will respond to its neighboring key.

Now we are ready to send a message. Suppose we want to send a telegram from New York to Philadelphia. The operator in New York opens his switch and presses down his key several times. The switch on the Philadelphia key being closed, the electricity goes through to the sounder, and, this being made an electromagnet by the current passing through the wire, the iron armature is attracted by the magnetism and drawn down to the magnet with a snap. It will stay there as long as the New York operator keeps his lever pressed down, but, when he allows it to spring up, there is no current passing through the Philadelphia sounder and there is no magnetism, consequently the armature springs up again with a click.

As often as the operator presses down his key lever and lets it spring up again, the same action takes place in the sounder, and it makes that click, click, which you have heard if you have ever seen telegraph instruments in operation.

Let us continue, however, to send our message. The New York operator, having pressed down his key several times to signal the Philadelphia operator, closes his switch to receive the answer from Philadelphia. The operator in the latter city then opens his switch and presses down his key several times, which makes the New York sounder click, in the same way, to let the operator there know that he is ready to receive the message. He then closes his switch and receives the telegram which the New York operator sends after opening *his* key.

Telegraphic messages are sent and received in this way and are read by the sound of the clicks.

These sounds may be represented on paper by dots, dashes, and spaces. For instance, if you press down the key and let it spring back quickly, that would represent a dot. If you press down the key and hold it a little longer before letting it spring up again, it would represent a dash. A space would be represented by waiting a little while before pressing down the key again.

We show you below the alphabet in these dots, dashes, and spaces, and these are the ones now used in sending all telegraphic messages.

A	B	C	D	E	F	G

H	I	J	K	L	M	N

O	P	Q	R	S	T	U

V	W	X	Y	Z		&

Thus, you see, if you were telegraphing the word "and" you would press down your key and let it return quickly, then press down again and return after a longer pause, which would give the letter A; then slowly and quickly, which would be N; then slowly and twice quickly, which would be D.

Any persevering boy can learn to operate a telegraph instrument by a little study and regular practice; and, as complete learner's sets can be purchased very cheaply, this affords a pleasant and useful recreation for boys.

There are many cases where two boys living near each other have a set of telegraph instruments in their homes and run a wire from one house to the other, thus affording many hours of pleasant and profitable amusement.

In giving the above explanation of telegraphing we have described only the simple and elementary form. In large telegraph lines, such as those of the Western Union, there are many more additional instruments used, which are very complicated and difficult to understand; such, for instance, as the quadruplex, by which four distinct messages can be sent over the same wire at the same time. We have, therefore, described only the simplest form in order to give the general idea of the working of the telegraph by electromagnetism, which is the principle of all telegraphing.

When you study electricity more deeply you will find this subject and the many different instruments very interesting and wonderful.

V

WIRELESS TELEGRAPHY

If it has seemed extraordinary to you that only one wire should be necessary for sending a message by the electric telegraph, and that our earth can be used instead of a second wire, how much more wonderful it is to realize that in these days we can exchange telegraphic messages with different points without any connecting wires at all between them, even though the places be many hundred miles apart. Thus, two ships on the ocean, entirely out of sight of each other, may intercommunicate, or may telegraph to or receive despatches from a far-distant shore; indeed, telegraphy without wires has been accomplished across the Atlantic Ocean. In the language of the day, this is called "wireless telegraphy," although it is more correct to think of it as aerial, or space, telegraphy. As you will naturally want to know how this is effected, we will try to explain the main principles in a simple manner.

If you drop a stone into a quiet pond, you will see the water form into ring-like waves, or ripples, which travel on and on until they die away in the far distance. These waves are caused, as we have seen, by a disturbance of the body of water.

Probably you have already learned in school that all known space is said to be filled with a medium called "ether," and that this medium is so exceedingly thin that it penetrates, or permeates, everything, so that it exists in the densest bodies as well as in free space. For the sake of obtaining a clear idea of this theory we may imagine that the ether envelops and permeates every thing in the entire universe. Hence we can easily realize that, although we cannot see or feel the ether, any disturbance of it will set it in wavelike motion.

Modern science accounts for light, radiant heat, and electrical phenomena by reason of wavelike disturbances, vibrations, or pulsations of this ether. Thus, if you should strike a light, the ether would be disturbed, causing waves to form, which, like the waves in the water, would travel in every direction. When these waves reached the eyes of another person within seeing distance, that person's eyes would be so acted upon by the waves that he would see the light which you had made, and would see it instantly, for light waves travel about 186,000 miles per second.

So, if you create an electrical disturbance, the same kind of an effect will be produced; that is to say, waves in the ether will be created, or propagated, and will travel on and on in every direction. Now, if some form of electrical appliance can be made that will be of the right kind to respond to them (as the eye responds to light rays), these electric waves can be made practically useful for transmitting messages through space. This is just what has been done, and we will now give you a brief general description of one kind of apparatus used.

For "sending," or "transmitting," as it is usually termed, there is used an induction-coil, having rather large brass balls on the secondary terminals; suitable batteries, a condenser, a Morse telegraph key, and an "aerial," or wire which is carried away up into the air vertically, and is made fast to a pole or special tower. When these are connected properly, the closing of the circuit with the key will cause sparks to jump between the brass balls. This electrical discharge, or oscillation, is carried by the aerial into the upper air and causes intense pulsations in the ether, which set up waves as already mentioned. If the circuit is opened again the disturbance ceases. So, by alternately closing and opening the circuit, the Morse characters can be imitated.

But how can these signals be received by the man for whom they are intended, who may be a hundred miles or more away? He has a "receiving" set, consisting of a sensitive relay, batteries, resistance-coils, a Morse register, an aerial, and a special device called a "coherer." This is the important part of the whole set, because it is sensitive to the electrical waves. It consists of a little glass tube about as large around as an ordinary lead-pencil, and perhaps two inches long. In the tube are two metallic plugs, each having a wire attached so that one wire projects from each end of the tube. The plugs are separated inside the tube by a very small space, and in this space are some metal filings. One wire from the coherer is connected to the aerial and the other to the ground. When there are no electrical ether waves to influence them, these filings, being loosely separated, are at rest and offer high resistance; but when the ether is disturbed by electrical vibrations and the waves arrive at the coherer (through the aerial), these filings are drawn together, or cohere. This lowers their resistance and they become a better conductor. Now, the coherer wires are also connected through a battery to the relay, which in turn is connected through another battery to a Morse register. Therefore, when the filings become a conductor, the current flows through them and the circuit to the relay is closed. That attracts an armature which closes the circuit of the Morse register and thus marks the electrical impulse on a strip of paper tape. In the mean time, a restoring device, called a "decoherer," operated also by the relay circuit, has tapped upon the coherer, thus shaking the filings loose again, so that they are ready to cohere again and register another impulse, or character. Thus, by pressing the key at the transmitting end for long or short periods, to represent Morse characters, long and short waves are propagated in the ether and are received and recorded at the receiving end through the coherer and other parts of the receiving set. In this way telegraphic messages are sent and received through space, between points separated by hundreds or thousands of miles.

We have tried to describe to you the general principles underlying the art of wireless telegraphy as plainly as possible, using for illustration the simplest kind of apparatus employed for the practical sending and receiving of messages. At the present day there are several systems in actual practice, and with the growth of the art there have been many elaborations of apparatus that have come into use. For instance, the coherer is not as much used as formerly. In its place there are employed several kinds of "wave-detectors" as they are now termed, and in many of the systems the electrical pulsations are generated by a dynamo-machine instead of batteries. Then, again, instead of the messages being recorded by a Morse register at the receiving end, the operator receives them by means of a telephone receiver, through which he hears the Morse characters and writes them down in words as he hears them. Generally the aerial, or "antennæ," as it is sometimes named, consists of several wires, sometimes a large number, carried to a considerable height.

There are a great many other details which might be written to explain all the complicated apparatus which is used in some of the systems, but it is not intended in this book to offer more than a general explanation of main principles. We must leave it to you to study the details elsewhere if you so desire after you have read these pages.

VI

THE TELEPHONE

You probably all know that the telephone is an electrical instrument by which one person may talk to another who is at a distance. Not only can we talk to a person who is in a different part of the city, but such great improvements have been made in these instruments that we can talk through the telephone to a person in another city, even though it be hundreds of miles away.

The main principle of the telephone is electromagnetism, as in the telegraph, but there are other important points in addition to those we mentioned in describing the latter.

Let us take first the

INDUCTION-COIL

You will remember that an electromagnet is made by winding many turns of wire around a piece of iron and sending a current of electricity through this wire.

Now, suppose this current of electricity was being supplied by two cells of a battery. If you took in your hands the wires coming from these *two cells*, giving, say, four volts, you could not feel any shock; but if you were to take hold of the ends of *the wires* on the *electromagnet* and *separate* them while this same current was going through, you would get a decided shock.

This separation would "break" the circuit, and the reason you would get a shock is that, while the electricity is acting on the wire, the iron itself is magnetized, and on breaking the circuit reacts upon the wire, producing for a moment more volts of pressure in every turn of it. Thus, you see, this weak pressure of electricity as it travels through the wire can yet produce, through its magnetism, strong momentary effects, but *you cannot feel it unless you break the circuit*.

HOW THE INDUCTION-COIL IS MADE

The object of the induction-coil is to produce high intensity, or pressure, from a comparatively weak pressure and large current of electricity; so, if we add still more wire, the magnet has a larger number of turns to act upon and thus makes a very strong pressure, or large number of volts, but a lesser number of ampères.

Instead of taking one piece of iron, as we would for an ordinary electromagnet, we take a bundle of iron wires in making an induction-coil, as these give a stronger effect. Around this bundle of wires we wrap many turns of insulated copper wire. This is called the *primary coil*, and the ends of this wire are to be attached to the battery.

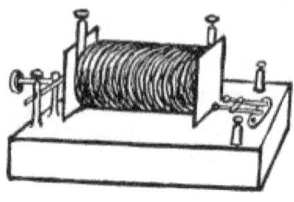

Fig. 9

On top of, or over, this primary coil we wrap a great many turns of very fine wire, of which, as it is so fine, a great length can be used. This is called the *secondary coil*, and it is in this coil that the volts, or pressure, of electricity become strongest.

Above we show you a sketch of an induction-coil. (Fig. 9.)

At the left-hand side of the cut is a "circuit-breaker," which is simply a piece of iron (armature) on a spring placed opposite the iron core. This armature is made a part of the wire leading to the primary coil. When the current from the battery is sent through the wires, the core becomes magnetized and draws this armature away from a fixed contact point, thus breaking the circuit, but the spring pulls it back, again completing the circuit, and so it keeps going back and forth very rapidly with a br-r-r-ing sound.

If you were now to take hold of the ends of the secondary coil you would get a continuous series of quick shocks which would feel like pins and needles running into you.

Perhaps most of you have taken hold of the handles of a medical battery and have had shocks therefrom. In so doing, you have simply had the current from the secondary of an induction-coil. The current may be made weaker by sliding a metallic cover over part of the iron core and so shutting off part of the magnetic effect.

SPARKING COILS

While on this subject we may add that these coils will produce sparks from the two ends of the wire of the secondary coil. These sparks vary in length according to the amount of wire in the coil. Small ones are made which give a spark a quarter of an inch in length, while others are made which will give sparks 10, 12, and 16 inches in length. In the latter, however, there are many miles of wire in the secondary coil.

The largest induction-coil known is one which was made for an English scientist. There are 341,850 turns, or 280 miles, of wire in the secondary coil. With 30 cells of Grove battery this coil will give a spark 42 inches in length. You may form some idea of the effect of this induction-coil when we state that if we desired to produce the same length of spark direct from batteries, without using an induction-coil, we should require the combined volts of pressure of 60,000 to 100,000 cells of battery.

Having explained to you briefly the induction-coil—how it is made and its action—we must ask you to bear these principles in mind, and presently we will tell you how it is used in the telephone.

The next thing we shall try to explain will be

THE VIBRATING DIAPHRAGM

Did you ever take the end of a cane in your hand, raise it up over your head, and then bring it down suddenly and sharply, so that it nearly touched the ground, as though you were about to strike something? If not, try it now with a thin walking-cane or with a pine stick about three feet long and one-half inch thick, and you will find that there is a peculiar sound given out. It is not the stick that makes this sound, but it is owing to the fact that you have caused the air to vibrate, or tremble, and thus give out a sound.

Fig. 10

If you strike a tuning-fork sharply you will see the ends vibrate and a sound will be given. If you put your fingers on top of a silk hat and speak near it you will feel vibrations of your voice.

Every time you speak you cause vibrations of the air; and the louder and higher you speak the greater the number of vibrations.

Suppose you take a thin piece of wood in your hands (say, for instance, the lid of a cigar-box cut in the shape shown in the picture, Fig. 10) and hold it about two inches from your mouth and then speak. You will feel the wood tremble in your hand. This is because the vibrations of the air cause the wood to vibrate in the same manner. These vibrations are very minute and cannot be seen with the naked eye, but they actually take place, and could be measured with a delicately balanced instrument.

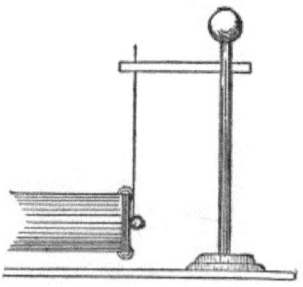

Fig. 11

Now let us try another experiment in further illustration of this principle. We will take a tube about three inches long and one and one-half or two inches in diameter. This tube may be made of cardboard. Now cut out a piece of thin cardboard which will just fit over one end of the tube. This piece we will call the "diaphragm." Fasten the diaphragm by pasting it with two strips of thin paper to the tube. These strips of paper should be fastened only on the ends, and the middle of the paper allowed to be slack, as shown in the picture, so that the diaphragm may work backward and forward easily. Take a small shot about the size seen in the sketch and tie it to a single thread of fine silk, then let it hang as shown in the sketch (Fig. 11), so that it will only just touch the diaphragm. Now, if you speak into the open end of the tube the diaphragm will vibrate and the shot will be seen to move to and from it according to the strength of the vibrations. If we could by any means make a diaphragm in another tube reproduce these same vibrations, we should hear the same words respoken, if the tube were held to the ear.

Fig. 12

While the vibrations caused by the human voice are too minute to be seen, it may seem surprising that they can be made to produce power. This is done by an ingenious mechanism called a Phonomotor, perfected by the great inventor Thomas A. Edison, of whom every one has probably heard. This mechanism, when spoken or sung at (or into) immediately responds by causing a wheel to revolve. No amount of blowing will start the wheel, but it can instantly be set in motion by the vibrations caused by sound.

The Phonomotor (which is shown in the engraving Fig. 12) has a diaphragm and mouthpiece. A spring, which is secured to the bedpiece, rests on a piece of rubber tubing placed against the diaphragm. This spring carries a pawl that acts on a ratchet or roughened wheel on the fly-wheel shaft. A sound made in the mouthpiece creates vibrations in the diaphragm; the vibrations of the diaphragm move the spring and pawl with the same impulses, and as the pawl thus moves back and forth on the ratchet-wheel it is made to revolve.

The instrument, therefore, is of great value for measuring the mechanical force of sound waves, or vibrations, produced by the human voice.

THE TRANSMITTER

That part of the telephone into which we speak is called the transmitter. This is usually a piece of hard rubber having a round mouthpiece cut through it. At the other side of this mouthpiece is placed a diaphragm made of a thin piece of metal, which is held m place by a light spring. Behind this diaphragm, and very close to it, is placed a carbon button. Between this carbon button and the diaphragm is a small piece of platinum, which is placed so as to touch both the button and diaphragm very lightly. This platinum contact piece is connected with one of the wires running to the primary of the induction-coil, and the spring attached to the carbon button is connected with the battery to which the other wire of the primary is connected. This is all shown in the sketch of a transmitter. (Fig. 13.)

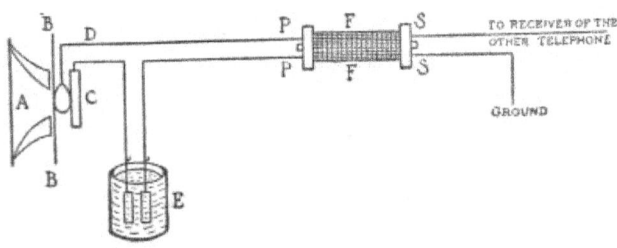

Fig. 13

A is the mouthpiece; B, the diaphragm; C, the carbon button; D, the wire at the end of which is the platinum contact; E, the battery; and F, the induction-coil; P, P are the wires to the primary, and S, S to the secondary wires.

We will now say a few words about the receiver, and then describe the manner in which the telephone works.

THE RECEIVER

This is that part of the telephone which is held to the ear, and by which we can hear the words spoken into the transmitter of the telephone at the other end of the line.

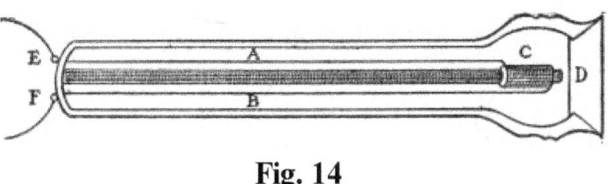

Fig. 14

The receiver is made of hard rubber, and contains a permanent bar magnet, which is wound with wire so as to make it also an electromagnet when desired. In front of this magnet is placed loosely a diaphragm of thin sheet iron. This diaphragm is placed so as to be within the influence of the magnet, but just so that neither one can touch the other.

Fig. 14 is a sketch of the receiver. A and B are the wires leading to the magnet, C, and D is the diaphragm. E and F are where the wires connect, one from the secondary of the induction-coil in the other telephone, and the other connected with the earth.

THE CARBON BUTTON

The little carbon button plays an important part in the telephone. You will see from the sketch of the transmitter that the current of electricity will flow through the carbon button to the contact point and through the wire to the primary of the induction-coil.

Now, carbon has a peculiarity, which is this, that if we press this carbon button, ever so slightly, against the platinum contact, there would be less resistance to the flow of the electricity through the wire to the primary, and the more we press it the less the resistance becomes. The consequence of this would be that more current would go to the primary, and the secondary would become correspondingly stronger. If the carbon button were left untouched, and nothing pressed against it, the flow of current through it would be perfectly even.

Having examined the inside of the transmitter and receiver, and understanding the effect of pressure on the carbon button, let us now see

HOW THE TELEPHONE WORKS

When we speak into the mouthpiece of the transmitter, the vibrations of the air cause the diaphragm to vibrate very rapidly, and, of course, every movement of the diaphragm presses *more or less* against the carbon button, in consequence of which the currents passing through the primary of the induction-coil are constantly increased or diminished and thus produce similar effects, but magnified, in the secondary.

The effect of this is that the magnet in the receiver of the other telephone is receiving a rapidly changing current, which, producing corresponding magnetic changes, makes the magnet alternately weaker or stronger. This influences, by magnetism, the iron diaphragm accordingly, and makes it reproduce the same vibrations that were caused by the speech at the transmitter of the sending telephone. Thus, the same vibrations being *reproduced*, the original sounds are given out, and we can hear what the person at the sending telephone is saying.

The action of the telephone illustrates well the wonderfully quick action of the electric current by the reproduction of these sound waves, or air vibrations, for they number many thousands in one minute's speech.

VII

ELECTRIC LIGHT

We have now arrived at a very interesting part of the study of electricity, as well as a more difficult part than we have yet told you of, but one which you can easily understand if you read carefully.

You must all have seen electric lights, either in the streets or in some large buildings, for so many electric lights are now used that there are very few people who have not seen them. But perhaps some of you have only seen the large, dazzling lights that are used in the streets, and do not know that there is another kind of electric light which is in a globe about the size and shape of a large pear, and gives about the same light as a good gas-jet.

These two kinds of electric lights have different names.

The large, dazzling lights which you see in the streets are called "arc-lights," and the small, pear-shaped lamps, which give a soft, steady light, are called "incandescent lights." We will tell you later why these names are given to them.

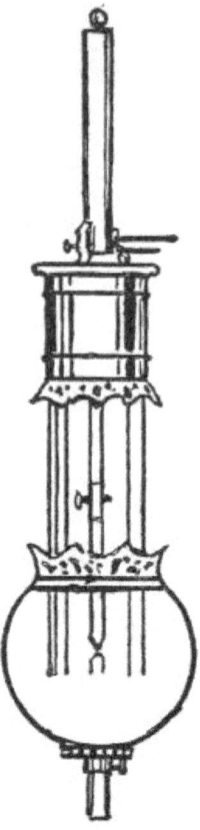

Fig. 15

The incandescent lights are generally used in houses, stores, theaters, factories, steamboats, and other places where a number of small lights are more pleasant to the eyes. The arc-lights (Fig. 15) are used to light streets and large spaces where a great quantity of light is wanted.

It would not be pleasant to have one of these dazzling arc-lamps in your parlor—although it would give a great deal of light—because your eyes would soon become tired. But two or three of the small incandescent lights (Fig. 16) would be very agreeable, because they would give you a nice, soft light to read or work by, and would not tire your eyes. So, you see, these two different kinds of lamps are very useful in their proper places.

Now, if you will read patiently and carefully, we will try and explain how both these lights are made.

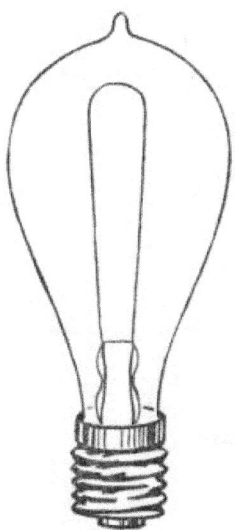

Fig. 16

You have seen that the telegraph, telephone, electric bells, etc., are worked by batteries. Electric lights, however, require such a large amount of current that it is too expensive to produce them in large quantities by batteries. A small number of lamps could be lighted by batteries, but if we were to attempt to use them to light 500 or 1,000 lamps together, the expense would be so enormous as to make it entirely out of the question.

There are many millions of incandescent lamps in use in the United States, but you will easily see that there could not be that number used if we had to depend on batteries to light them. You will understand this more thoroughly when you have finished reading this little book.

Well, you will ask, if we cannot use batteries, what is used to produce these electric lights?

Machines called "dynamo-electric machines," or "generators," which are driven by steam-engines or water-power, are used to produce the electricity which makes these lamps give us light.

You will remember that in the chapter on Magnetism we explained to you how electricity makes magnetism, and now we will explain how, in the dynamo, magnetism makes electricity.

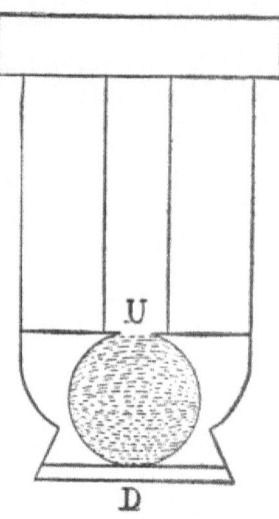

Fig. 17

It has been found that the influence of a magnet is very strong at its poles, and that this influence is always in the same lines. This influence has been described as "lines of force," which you will see represented in the sketch above by the dotted lines (Fig. 17). Of course, these lines of force are only imaginary and cannot be seen in any magnet, but they are always present. The meaning of this term "lines of force," then, is used to designate the strength of the magnet.

Many years ago the great scientist Faraday made the discovery that, by passing a closed loop of wire through the magnetic lines of force existing between the poles of a magnet, the magnetism produced the peculiar effect of creating a current of electricity in the wire. If the closed loop of wire were passed down, say from U to D, the current flowed in the wire in one direction, and if it were passed upward, from D to U, the current flowed in the other direction. Thus, you see, magnetism produces electricity in the closed loop of wire as it cuts through the magnetic lines of force. Just why or how, nobody knows; we only know that electricity is produced in that way, and to-day we make practical use of this method of producing it by embodying this principle in dynamo-machines, as we will shortly explain.

In carrying this discovery into practice in making dynamo-machines we use copper wire. If iron were used, there would be a current of electricity generated, but it would be much less in quantity, because iron wire has much greater resistance to the passage of electricity than the same size of copper wire.

Perhaps you can understand it more thoroughly if we state that when a closed loop of wire is passed up and down between the poles of a strong magnet there is a very perceptible opposition felt to the passage of the wire to and fro.

This is due to the influence of the magnetism upon the current produced in the wire as it cuts through the lines of force, and, inasmuch as these lines of force are always present at the poles of a magnet, you will see that, no matter how many times you pass the loop of wire up and down, there will be created in it a current of electricity by its passage through the lines of force.

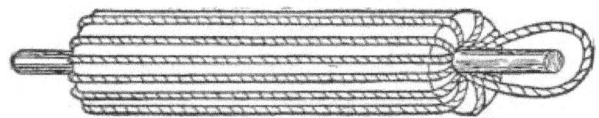

Fig. 18

Suppose that, instead of using one single loop of copper wire, you wound upon a spool a long piece of wire like that in Fig. 18, and that you turned this spool around rapidly between the poles of the magnet, you would thus be cutting the lines of force by the same wire a great many times, and every time one length of the wire cut through the lines of force some electricity would be generated in it, and this would continue as long as the spool was revolved. But, as each length would only be a part of the one piece of wire, you will easily see that there would be a great deal of electricity generated in the whole piece of wire.

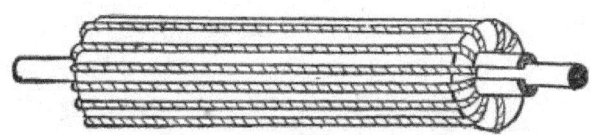

Fig. 19

All we have to do, then, is to collect this electricity from the two ends of the wire, and use it. If we should attach two wires to the two ends of this wire on the spool, they would be broken off when it turned around, so we must use some other method. We fix on the end of the spool (which is called an "armature") two pieces of copper, so that they will not touch each other (as in Fig. 19), and fasten the ends of the wire to these pieces of copper. This is called a "commutator," and, as you see, is really the ends of the wire on the spool. Now we get two thin, flat pieces of copper and fix them so that they will rest upon the copper bars of the commutator, but will not go round with it. These two flat pieces of copper are called the "brushes," and they will collect from the commutator the electricity which is gathered in the wire around the spool. As the brushes stand still, two wires can be fastened to them, and thus the ampères of current of electricity, acted upon by the volts pressure, can be carried away to be used in the lamps, for you must remember that as long as the spool turns around it gathers more electricity while there is any magnetism for the wire on the spool to pass through. The constant revolving of the spool creates so much electricity that it is driven out from the wire on the spool, through the commutator to the brushes, and there it finds a path to travel away from the pressure of the new electricity which is all the time being made.

In this way we get a continuous current of electricity in the two wires leading from the commutator, and can use it to light electric lamps or for other useful purposes.

In explaining this to you, so far, we have used as an illustration of the magnet one of the steel permanent magnets in order to make the explanation more simple, but now that you understand how the electricity is made, we must explain to you something about the magnets that are used in dynamo-machines. We can perhaps make this more clear by giving another example.

Suppose you had a dynamo which was lighting up 100 of the incandescent lamps, each of 200 ohms resistance and each requiring 100 volts pressure. Now each lamp would take just a certain quantity of electricity, say half an ampère; so, the 100 lamps would require one hundred times that quantity. But, if you turned off 50 of these lamps at once, the tendency would be for the pressure to rise above the 100 volts required for the other 50, and they would be apt to burn out quicker. It is plainly to be seen, then, that we must have some means of regulating the magnetism so as to regulate the lines of force for the wire on the armature to cut through. We can do this with an electromagnet, but not with a permanent magnet, because *we cannot easily regulate the amount of magnetism which a permanent magnet will give.*

There is another reason why we cannot use permanent magnets in a dynamo, and that is because *they cannot be made to give as much magnetism as an electromagnet will give.*

Thus you will see that there are very good reasons for using electromagnets in making dynamo-machines. Let us see now how these electromagnets and dynamos are made, and then examine the methods which are followed to operate and use them.

You must remember, to begin with, that in referring to wire used on magnets and armatures and for carrying the electricity away to the lamps, we always mean wire that is *covered* or *insulated.* In electric lighting, insulated wire is *always* used, except at the points where it is connected with, the dynamo, the lamps, a switch, or any point where we make what is called a "connection."

As the shape of the magnets is different in the dynamos of various inventors, we will take for illustration the one that is nearest the shape of the horseshoe and the shape that is generally used in illustrating the principle of the dynamo. This is the form used by Mr. Edison, whom we have previously mentioned. This form is shown in Fig. 20.

Now, although this magnet appears to be in one piece, it really consists of five parts screwed together so as to make, practically, one piece. The names of the parts are as follows: F, F are the "cores"; C the "yoke," which binds them together; and P, P the "pole pieces," where the magnetism is the strongest. These pole pieces are rounded out to receive the *armature*, which, as you will remember, is the part that turns around.

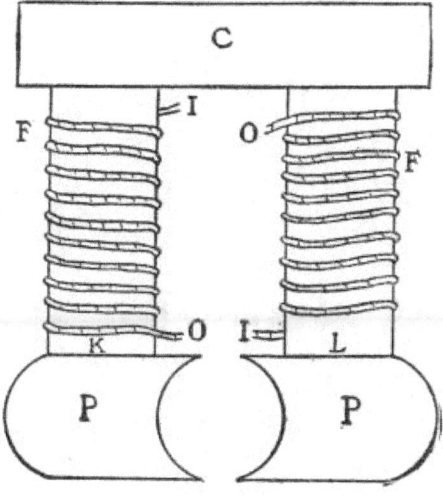

Fig. 20

The cores, F, F, are first wound with a certain amount of wire, which depends upon the use the dynamo is to be made for. Thus, you will see, there will be on each core two loose ends of the wire that is wound around it—namely, the beginning of the wire and the end where we leave off winding, which on the two cores together will make four ends of wire. We will tell you presently what is done with them.

After the cores are wound, they are screwed firmly to the yoke and to the pole pieces, so as to make, for all practical purposes, one whole piece pretty nearly the shape of a horseshoe magnet.

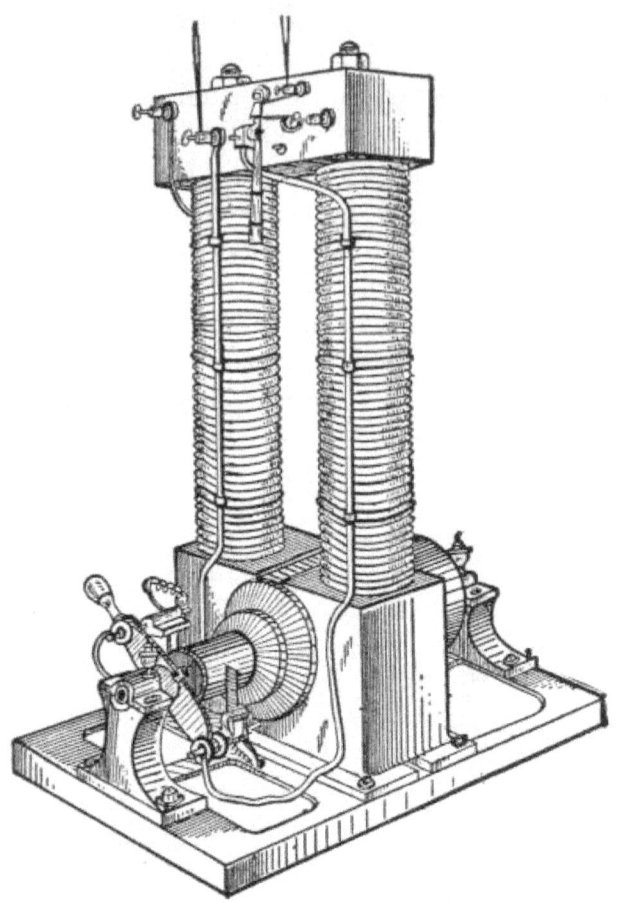

Fig. 21

Now, to make the dynamo complete, we must put in the armature between the poles, which are rounded off, as you will see, to accommodate it. The armature is held up by two "bearings," which you will see in the sketch of the complete dynamo above. (Fig. 21.)

The armature in a practical dynamo-machine consists of a large spool made of thin sheets of iron firmly fastened together and having a steel shaft run through the center, upon which it revolves.

This spool, or armature, is wound with a number of strands of copper wire. The commutator, instead of consisting of two bars, is made in many dynamos with as many bars as there are strands of wire, and the ends of these wires are fastened to the bars of the commutator so as to make, practically, one long piece of wire, just as we showed you in explaining how the electricity was produced.

The brushes, resting upon the commutator, carry away the electricity from it into the wires with which they are connected.

Now we have our dynamo all put together and ready to start as soon as we properly connect these four loose ends of wire on the cores.

If you will turn back to Fig. 20 you will see that two of the wires are marked I, and the other two O. The letter I means the inside wire, or where the winding began, and the letter O means the outside wire, or where we left off winding.

Now, if we fasten together (or "connect") the two ends of wires, I and O, near the top of the magnet, we make the two wires round the cores into one wire, which starts, say, at I near the poles, goes all around one core, crosses over and around the other core down to the other end of the wire to O, near the poles.

So far we have called the iron a magnet, although it is not a magnet until electricity is put into it; so, when the dynamo is started for the first time, these two ends of wire, I and O, are connected to a battery or other source of current for the purpose of sending electricity through the wire on the cores. When the electricity goes into this wire the iron immediately becomes a magnet, and the lines of force are present at the poles.

Now, the armature is turned around rapidly by a steam-engine, and, as the wire on the armature cuts the lines of force with great rapidity and so frequently, there is quickly generated a large quantity of electricity, which passes out as fast as it is made through the commutator and the brushes to the lamp. And so long as the armature is revolved and the battery attached, the electricity will be made, or, as it is usually termed, "generated."

As we stated above, a battery is used *the first time the dynamo is run*, and now we will explain why it is not needed afterward.

Although iron will not become a permanent magnet, like steel, it *does not lose all its magnetism* after it has been once thoroughly charged. When the dynamo is stopped, after the first trial, and the battery is taken away, you will discover only traces of magnetism about the poles. They will not readily attract even a needle or iron filings; but there is, nevertheless, a very small amount of magnetism left in the iron. Small as this magnetism is, however, it is enough to make very faint and weak lines of force at the poles of the magnet.

After the battery is taken away, the ends of the wire on the cores, which were connected to the battery, are connected, instead, to the wires which carry away the electricity from the brushes to the lamps. Thus, you will see, if any electricity goes from the dynamo to the lamps, part of it must also find its way through the wires which are around the cores.

We will now start up the dynamo without having any battery attached and see what happens. The armature turns around and the wires upon it cut through those very faint lines of force which are always at the poles. This, as you know, makes some electricity; very little, to be sure, but it comes out through the brushes to the wires leading to the lamps, and there it finds the wires leading back to the cores. Well, part of this weak current of electricity goes into these wires and travels back round the cores and so makes the magnetism stronger. The consequence of this is that the lines of force become stronger and, as the armature keeps turning around, the electricity naturally becomes stronger, and so there is more of it going through the wires back to the cores and increasing the strength of the magnet all the time, until the dynamo becomes strong enough to generate all the current it was intended to give for the lamps.

Of course, you understand that the stronger the magnet becomes, the greater will be the lines of force and the greater the amount of electricity made by the turning of the armature. Now, there is naturally a limit to what can be done with any particular dynamo; so, while the electricity continues to strengthen the magnetism and the magnetism increases the electricity, this cannot go beyond what is called the "saturation" point of the magnet.

Saturation means that the iron is full of magnetism, and will hold that much but no more. You will learn more as to the saturation of magnets when you study electricity more deeply, and we therefore do not intend to enter into that subject in this book. We will only state, however, that the magnets of dynamos are not always charged up to their saturation point.

THE LAMPS

So far you have learned how the current of electricity is produced, and now we will follow along the wires to find out how it makes the lamps give out both strong lights and the smaller, pleasant ones.

Suppose we take first the large, dazzling lights we see in the streets, which, as you know, are called

ARC-LIGHTS

Those who have seen the arc-lamps will readily recognize them from the picture in Fig. 22.

You will see that there are two sticks, or "pencils," of carbon. Now you will remember that in the chapter on Magnetism we told you that *in order to have electricity do work for us we must put some resistance or opposition in its way.* When we get light from an electric lamp it is because we make the electricity do some work in the lamp, and this work is in pushing its way through a resistance or opposition which is in the lamp.

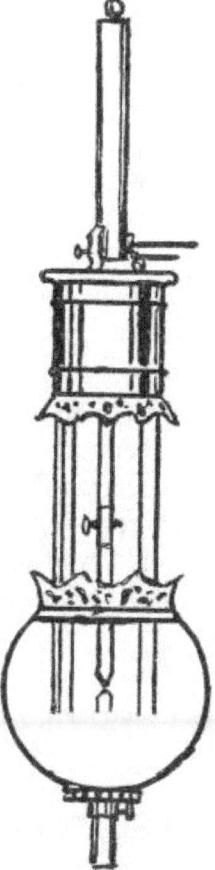

Fig. 22

When we generate electricity in the dynamo and put two wires for it to travel in, the current goes away from the dynamo through one of the wires and will go back to the dynamo through the other one if it can possibly get a chance to get to this other one. Now, the electricity which is constantly being made fills the wires and acts as a pressure to force the current through the wires back to the dynamo, and, if we put no resistance or opposition in the way, it would have a very easy path to travel in and would do no work at all. The wires leading to an electric lamp should have very little resistance, not sufficient to require any work from the current in passing through.

So, if we bring the two carbons in an arc-lamp together they really form part of the wire, and do not interrupt the current in its travels, but, if we *separate the carbons*, we make a gap which the current must jump across if it wants to go on. As the volts, or pressure, is so great, the current must jump, and this *against the resistance or opposition* in an arc-lamp is that which gives the current so much work to do. Indeed, so hard is it for the current to jump across this gap that it breaks off from one carbon a shower of tiny particles as fine as the finest dust, and makes them white hot in passing to the other. This shower of fine carbon dust, together with the ends of the carbons, being white hot, of course makes a light, and this is the dazzling light which you see in the arc-lamp.

Of course, when the electricity has jumped over from one carbon to the other, it goes through it to the wire, and so passes on to the next lamp, where it has to jump again, and so on until it has gone through the last lamp, then it has an easy path to get back to the dynamo.

Now, we want you to understand more thoroughly how that much resistance or opposition will cause heat, so we will try to give you a simple example.

Most of you know that if you were holding a rope tightly in your hands and some one pulled it through them quickly and suddenly, it would get very hot and your hands would feel as though they were being burned. This is heat caused by your hands resisting or opposing the passage of the rope through them, and if you could hold on tightly enough and the rope was drawn through quickly enough, it would take fire. This fire would, therefore, cause heat and light.

It is just this principle of resistance to the passage of the current which causes the light in an arc-lamp, as we have shown you.

INCANDESCENT LAMPS

You have just learned that the light in an arc-lamp is caused by the current forcing off from the carbon sticks tiny particles and heating them up until they give a brilliant light. So, you see, in an arc-light there is a wearing away of carbon by electricity, and therefore these sticks, or pencils, of carbon in time are all burned away. In practice the carbon pencils last about eight or ten hours, and then new ones must be put in.

Now, in the incandescent lamp there is also carbon used, but the light is not produced by the combustion or wasting away of the carbon, as we will show you.

The picture below will show you the appearance of an incandescent lamp. (Fig. 23.)

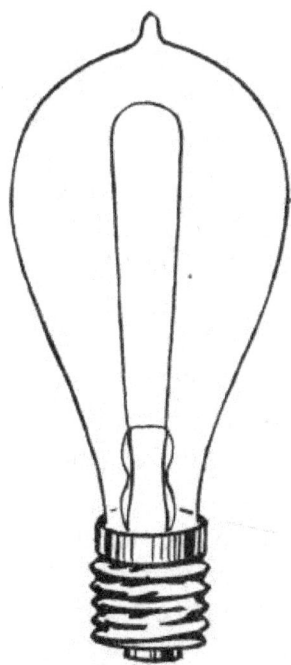

Fig. 23

You will see that this lamp consists of a pear-shaped globe, and inside is a long U-shaped strip of carbon no thicker than an ordinary thread. This is a strip of bamboo cane[1] which has been carbonized to a thread of charcoal. It is joined to two wires which come through the glass. These two wires come down through the bottom of the globe, and one is fastened to a brass screw-ring, while the other wire is fastened to a brass button at the bottom of the lamp. These two (the ring and button) must, as you know, be separated from each other by something which will not carry electricity, or they would make a short circuit when the electricity was applied. We separate the ring and the button in various ways.

Now, if we took the ends of two wires which were charged with the proper amount of electricity and put one wire on the screw-ring and the other on the button, the lamp would light up, because there would be a complete path for the current to travel in.

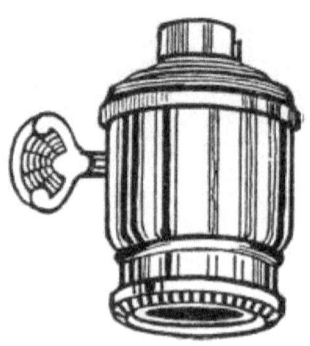

Fig. 24

It will, however, be plain to you that it would be awkward to light the lamps in this way, so we use a "socket" into which the lamp is screwed. (Fig. 24.)

The wires from the dynamo carrying the electricity are connected in the socket, one wire with the screw thread into which the screw-ring fits, and the other with a button which the button on the lamp touches when the lamp is screwed into the socket. Thus we have a connected path for the current to travel in, or, as it is termed, a *complete circuit*.

You will notice that in the incandescent lamp the electricity does not need to jump, as it does in the arc-light, because we give it one continuous line to travel in.

In order, however, to get the current to do work for us, we put some resistance in its path, which it must overcome in order to travel back to the dynamo. The resistance in an incandescent lamp is the U-shaped carbon strip (or, as it is called, "filament"). This charcoal filament has so much greater resistance than the wires that it opposes, or resists, the passage of the electricity through it; but the electricity *must* go through, and, as it is strong enough to force its way, it overcomes this resistance and passes on through the carbon to the wire at the other end. You see it is a struggle between the carbon and the electricity, the current being determined to go on and the carbon trying to keep it back; and, in the end, the electricity, being the stronger, gets the best of it; but the struggle has been so hard that the carbon has been raised to a white heat, or incandescence, and so gives out a beautiful light, which continues as long as the current of electricity flows.

You will remember that in the arc-light the carbons are slowly consumed and new ones must be put in. If the carbon in the incandescent light were consumed, it would not last many minutes, because it is only about the size of a horsehair. Now, you will naturally inquire why this fine strip is not burned up when it is raised to so high a heat. Well, we will tell you.

You know that if you light a match and let it burn the wood will all be consumed. But did you ever light a match, put it into a small bottle, and put the cork in? If you never did, do so now as an experiment, and you will see that the match will keep lighted for an instant and then go out without consuming the wood.

The reasons for this are very simple. In order to burn anything up entirely it is absolutely necessary to have the gas called oxygen present, and, as the air you live in contains a very large amount of oxygen, there is more than sufficient in your room to cause the wood of the match to be entirely consumed after it is lighted. But there is such a small quantity of oxygen in the bottle that it is not enough to keep the fire going in the match, and, consequently, it will not burn up the wood.

The reason the filament in an incandescent lamp is not burned up is because there is *no oxygen* inside the globe. After the carbon is put in its place all the oxygen is drawn out through a tube, and the glass is sealed up so that no more oxygen can get in. This is called obtaining a "vacuum," and vacuum means a space without air.

There being no oxygen in the globe, it is impossible for the carbon to burn up; so the incandescent lamp will continue to give its light for a very long time, some of them lasting for thousands of hours. Some day, however, from a great variety of obscure causes, the filament becomes weak in some particular spot and breaks, and the light ceases. When this happens, we unscrew the lamp and put another one in, and the light goes on as usual.

Now you have learned how the incandescent lamp is made to give light. We will add that it is a beautiful, soft, white light, almost without heat, it will not explode, throws off no poisonous fumes like gas or oil lamps, and has many other points of comfort and convenience which make it very desirable.

ELECTRIC-LIGHT WIRES

Before closing the subject of electric light you would perhaps like to know something about the way in which we place the wires leading to the lamps.

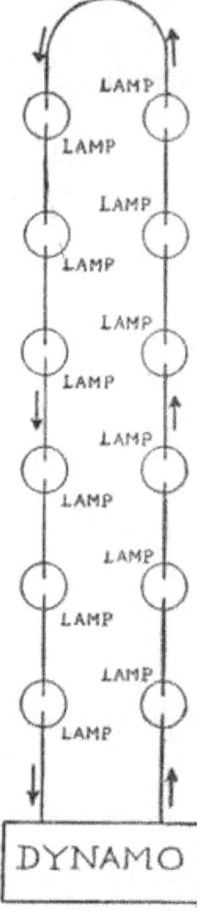

Fig. 25

If you remember what we told you about measurements in the beginning of this book, it will be easy to understand what follows:

You know that if you have a very great pressure you can force a quantity through a small conductor. This is the principle upon which the arc-lamps are run. Every arc-lamp takes about 40 to 50 volts and from 5 to 10 ampères to produce the light, and they are connected with the wires as shown in Fig. 25.

This is called running lamps in "series," and, as you will see from the sketch, the wire starts out from the dynamo and connects with one carbon of the first arc-lamp, and to the other carbon is connected another wire which goes on to the next lamp, and so on until the last lamp is reached, and then the wire goes back to the dynamo. This forms, practically, one continuous loop from one brush to the other of the dynamo.

The current starts out, makes its way through the first lamp, goes on to the next, makes its way through that, and so on till it has jumped the last one; then it goes back to the dynamo.

Now, as each of these jumps requires a pressure of 40 or 50 volts, you will easily see that the total pressure, in volts, of the electricity must be as many times 40 or 50 volts as there are lamps to be lighted; so, if there were 60 lamps in circuit, there would be 2,400 to 3,000 volts pressure, which, while it gives very fine lights, might cause instant death to any one touching the wires.

Suppose anything happened to the first lamp, which stopped the current from jumping through it. There would be no path for the current to travel farther, and, consequently, all the lights would go out. To get over this difficulty there is sometimes used what is called a "shunt," which only acts when the lamp will not light. This shunt carries the current round the lamp to the other wire, so that it may travel on and light up the other lamps.

WIRES FOR INCANDESCENT LAMPS

The wiring for incandescent lamps is carried out in an entirely different way, which you can see by comparing Fig. 25 A with Fig. 25 which shows the wiring for arc-lamps.

25 A

This is called connecting in "multiple arc."

You will notice that the two wires running out from the dynamo (which are called the main wires) do not form one continuous loop as in the arc-light system, but that a smaller wire is attached to one of the main wires and then connected with the screw-ring in the lamp-socket; then another wire is connected with the button in the socket and afterward to the other main wire. Every lamp forms an independent path through which the current can travel back to the dynamo.

Now, if we turn one of these incandescent lamps out, we simply shut off one of these paths and the electricity travels through the other lamps, and, if we wish, we can turn out all the lamps but one and there will still be a way for the electricity to go back to the dynamo.

In the arc-lamps we must have a very high number of volts pressure, because the electricity has only one path, and it all has to pass through the first and other lamps till it comes to the last one. In the incandescent light the electricity has as many paths as there are lamps, so we only need to keep *one* certain *pressure* in volts in the main wires all the time. This pressure is *even* all the way through the main wires, and, therefore, it is ready to light a lamp the instant it is turned on, because, as you have seen, electricity will always get back to the dynamo if there is a possible chance, and the lamp opens a path.

The volts pressure used to operate any number of incandescent lamps is altogether very much less than for a number of arc-lights. For example, in the Edison system the pressure (sometimes called "electromotive force") is only about 110 volts, which is very mild and not at all dangerous. This electromotive force would be *the same* if there were *one lamp or ten thousand* lighted.

While this Edison current would not hurt any one, you should remember that it is much the better plan not to touch *any* electric-light wires until you have learned a great deal more on this subject.

We may add that each of the standard incandescent lamps requires only about one-quarter of an ampère of current to make them give a light of 16 candle-power, which is about the light given by a very good gas-jet, and while the electromotive force, or pressure, would only be about 110 volts, whether there were one lamp or ten thousand lighted, there must be sufficient ampères in the wires to give each lamp its proper quantity.

SWITCHES

We have made mention several times of turning on or off one or more lights, and now, perhaps, you would like to know how this is done.

Suppose the electricity was traveling through wires to one or several lamps, it would light up those lamps as long as the wires provided a path to travel in, but if you were to cut out one of them, which is called "breaking the circuit," there would be no road for the electricity to follow, and, consequently, its course would be stopped short and the lamps would go out. You will remember that *electricity must have a complete circuit* or it can do no work, and in electric lighting it is always a *metallic circuit* that is used.

Now, the switch is simply a device which is used to break the circuit so that the current cannot pass on. The simplest form of switch is seen in the sketch. (Fig. 26.)

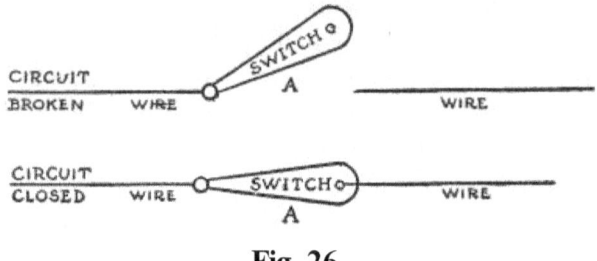

Fig. 26

You will see that there is a wire cut in two, and to one piece is attached a metallic piece, A, which turns one way or the other, and when it is turned so as to touch the other part of the wire the circuit is closed and the electricity goes from the lower part of the wire through the metallic piece A to the other part of the wire, thus making a complete circuit or path for the electricity to travel in.

If we turn the piece A away from the upper wire this breaks the circuit and cuts off the path, and, of course, the lamps would go out.

This is the principle of the switch, and, although they are made in thousands of ways, switches all have the same object—namely, the closing and breaking of the circuit, whether it is for one or a hundred lamps.

WIRE ON DYNAMOS

In explaining to you the construction and working of dynamo-machines, we did not state anything about the amounts of wire used in winding the machine.

It is not our intention to say exactly how much is used on any one dynamo, because that is among the things you will have to learn when you come to study the subject of electricity more deeply.

We simply want to have you understand that upon the number of turns of wire on any one machine depends the effect that that amount of wire, carrying electricity, will have upon a certain weight of iron when the armature is revolved a certain number of turns per minute.

A certain number of strands of wire on an armature will only do a certain amount of work at the most, so you will see that a small dynamo will not produce as much electricity as a larger one containing more iron and wire. For high pressure there must be more strands of wire cutting the lines of force more frequently than would be required for low pressure; and, to produce a great many ampères, the armature must be larger and the wire upon it thicker than it would need to be if only a small number of ampères were wanted.

This of itself is a very deep and complicated subject, and many books have been written upon it alone. We shall, therefore, not attempt to go more deeply into it in this little book, but simply content ourselves with giving you the general idea, which will be sufficient until you make a thorough study of the subject.

VIII

ELECTRIC POWER

One of the most convenient uses to which electricity is put is in producing motive power for driving all kinds of machines, from a sewing-machine to a railway train, and we will now try to explain how we can get this kind of work from electricity.

To begin with, you all know that a piece of machinery is usually made to work by revolving a wheel which is part of the machine, either by means of a steam-engine or by water-power, or, as a sewing-machine, by foot-power. Now, when we work a piece of machinery by electricity we do just the same thing by using, instead of the steam-engine or water or foot power, an electric-engine called an "electromotor," which operates in the same way—namely, by turning the wheel of the machine it is applied to.

Foot-power is hard work for the person who is applying the power, and, as you can easily see, one person can make only a very little power by use of the feet. Steam and water power can be used for any large amount of work, but the work must be within a few hundred feet of the engine or the power cannot be used.

If there were a factory using steam-power a block or two away from where you lived, and you had a lathe in your house which you would like to have run by the steam-power in the factory, it would be practically impossible to do this. Now, if the factory were still farther away from your house, it would be still more impossible, and if it were a mile away it would be foolish to dream of taking steam-power from a place so far away.

Suppose, however, that this factory was lighted by electric lights, it would be a very easy matter to take some of the power over to your house. This could be done, even if the factory were miles away, by taking two wires from their electric-light wires and running them into your house to an electromotor connected with your lathe. This electromotor would then run your lathe just as well as if it were belted to a steam-engine.

So, you see, power can be carried in the form of electricity through two wires over very great distances and made to do work at a long way from the engine which is turning the dynamo to make the electricity. Thus, you may have brought into your house wires which will give lights and, at the same time, power to run a sewing-machine, a lathe, or any other piece of machinery.

Having learned so far that a dynamo will make a continuous current of electricity, and that two wires will carry this current to any place where it is wanted, let us now see what takes place in the electromotor to transform the electricity into power.

An electromotor (which we will now call by its short name, motor) is simply a machine made like a dynamo. Curious as it may seem to you, it is a fact that if you take two dynamo-machines exactly alike, and run one with the steam-engine so as to produce electricity, and then take the two main wires and attach them to the brushes of the other dynamo, the electricity will drive this other dynamo so as to produce a great deal of power which could be used for driving other machines. Thus, the second dynamo would become a motor.

In the chapter on dynamos we explained something about the way they were made and how the electricity was produced.

THE MOTOR

You will remember that the armature consists of a spool wound with wire. This spool is made of iron plates fastened together so as to form one solid piece. The armature of a motor may be made in the same way; in fact, the whole motor is practically a dynamo-machine.

There is something more about magnetism which we will tell you of here, because you will more easily understand it in its relation to an electromotor.

If we take an ordinary piece of iron and bring one end of it near to (but not touching) one pole of a magnet, this piece of iron will itself become a weaker magnet as long as it remains in this position. This is said to be magnetism by "induction." The end of the piece of iron nearest to the magnet will be of the opposite polarity. For instance, if the pole of the magnet were north, the end of the iron which was nearest to this north pole would be south, and, of course, the other end would be north. To make this more plain we show it in the following sketch. (Fig. 27.)

This would be the same whether the magnet were a permanent or an electromagnet.

You will remember also that the north pole of one magnet will *attract the south pole* of another magnet, but will *repel a north pole*.

These are the principles made use of in an electromotor, and we will now try to show you how this is carried into practice.

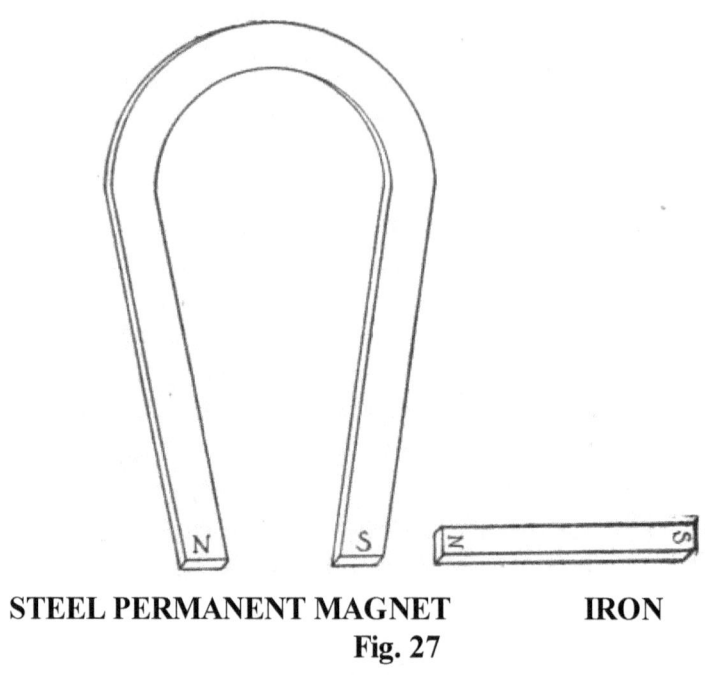

STEEL PERMANENT MAGNET   **IRON**
Fig. 27

Although a motor is made like a dynamo, we will show a different form of machine from the dynamo already illustrated, because it will help you to understand more easily. (Fig. 28.)

Here we have an electromagnet with its poles, and an iron armature wound with wire, just as in the dynamo we have described, except that its form is different.

Fig. 28

A commutator and brushes are also used, but the electricity, instead of being taken away from the brushes, is taken *to* them by the wires connected with them. Two wires are also connected which take part of the electricity around the magnet, just as in the dynamo.

Now, when the volts pressure and ampères of electricity coming from a dynamo or battery are turned into the wires leading to the brushes of the motor, they go through the commutator into the armature and round the magnet, and so create the lines of force at the poles and magnetize the iron of the armature.

Let us see what the effect of this is.

The poles of the magnet become north and south, and the four ends on the armature also become north and south, two of each.

By referring to Fig. 28 again we shall see what takes place.

The north pole of the magnet is doing two things: it is repelling, or forcing away, the upper north pole of the armature and at the same time drawing toward itself the lower south pole of the armature.

In the mean time the south pole of the magnet is repelling the south pole of the armature and at the same time drawing toward itself the north pole of the armature.

This, of course, makes the armature turn around, and the same poles are again presented to the magnet, when they are acted upon in the same manner, which makes the armature revolve again, and this action continues as long as electricity is brought through the wires to the brushes. Thus, the armature turns around with great speed and strength, and will then drive a machine to which it is attached.

The speed and strength of the motor are regulated by the amount of iron and wire upon it, and by the volts pressure and ampères of electricity supplied to the brushes. Motors are made from a small size that will run a sewing-machine up to a size large enough to run a railway train, and are often operated through wires at a great distance from the place where the electricity is being made, sometimes miles away.

They are also made in a great many different forms, but the principle is practically the same as we have just described to you.

IX

BATTERIES

So far we have only described one way of producing electricity—namely, by means of a dynamo-machine driven by steam or water power. The supply of electricity so obtained is regular and constant as long as the steam or water power is applied to the dynamo.

There is another and very different way of producing electricity, and this is by means of a chemical process in what is called a battery.

To obtain electricity from the dynamo we must spend money for the coal to make the steam which operates the steam-engine, or for the water which turns the water-wheel, as well as for an engineer in both cases. When we obtain electricity from a battery we must spend money for the chemicals and metals which are constantly consumed in the battery.

PRIMARY BATTERIES

An electrical battery is a device in which one or more chemical substances act upon a metal and a carbon, or upon two different metals, producing thereby a current of electricity, which will continue as long as there is any action of the chemicals upon the metal and carbon, or upon the two metals.

Batteries for *producing* electricity may be divided into two classes, called "open circuit" batteries and "closed circuit" batteries.

Open-circuit batteries are those which are used where the electricity is *not* required constantly without intermission—for instance, in telephones, electric bells, burglar alarms, gas-lighting, annunciators, etc.

Closed-circuit batteries are those which are used where the effect produced must be continuous every moment, as, for instance, in electric lights and motors.

The open-circuit battery is made in many different ways, so we only describe two of the principal ones.

As we told you in an early part of this book, we do not know just what electricity is, nor why it is produced under the conditions existing in a battery. But we do know that by following certain processes and making certain chemical combinations we can make as much electricity and in such proportions as we want.

The two metals, or the metal and carbon, in a battery are called the "elements," and to these are connected the wires which lead from the battery to the instruments to be worked by it.

The Leclanché Battery.—This form of open-circuit battery consists of a glass jar in which is placed the elements. One element consists of a rod of zinc, and the other element is carbon and powdered black oxide of manganese. These two (the carbon and black oxide of manganese) are placed in an earthenware vessel called a "porous cup." This is simply a small jar made of clay which is not glazed. Thus, the liquid which is in the glass jar penetrates through the porous cup to the carbon and manganese which it contains, and so the chemicals affect both these and the zinc at once, for, in order to obtain electricity, you will remember that the chemical action must take place at the same time upon both the elements in the same vessel. (Fig. 29.)

The chemical substance used in this battery is sal-ammoniac, or salts of ammonia. A certain quantity of this salt is dissolved in water, and this solution is poured into the glass jar. When this is done the battery will generate electricity at once.

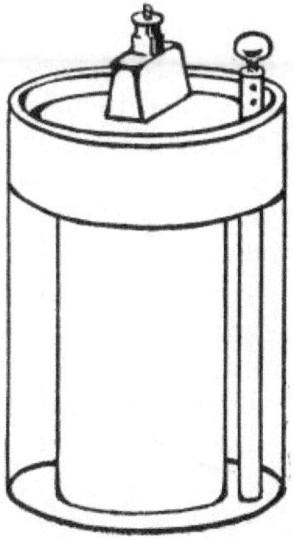

Fig. 29

It should be remembered that the proper term for the chemical mixture which acts upon the elements in any battery is "electrolyte."

The Dry Battery.—The cleanliness, convenience, high efficiency, and comparatively low internal resistance of the dry cell has brought it into great favor in the last few years. It is now extensively used in preference to the Leclanché and other open-circuit batteries having liquid electrolyte for light work, such as bells, gas-lighting, burglar alarms, ignition on motor-boats, automobiles, etc.

The dry cell is also used in great numbers for pocket flash-lamps, and in other ways where it would be impossible to employ batteries containing liquids.

A dry cell consists of zinc, carbon, and the electrolyte, which is a mixture so made that it is in the form of a gelatinous or semi-solid mass, so that it will not run or slop over.

A piece of sheet zinc is formed into a long tube, and a round, flat piece of zinc is soldered at one end, thus making a cup open at one end. This forms the cell itself, and at the same time becomes one of the elements. The other element is a piece of battery carbon which is long enough to project out of the top of the cell about half an inch or more. While the cell is being filled with the electrolyte the carbon is held up by a support so that it does not touch the zinc at the bottom of the cup. Of course, the zinc cup and the carbon are provided with proper binding-posts or other attachments, so that conducting wires can be connected.

The electrolyte is packed into the cup and around the carbon in such a way that the cup is entirely filled within about half an inch from the top, and then some melted tar or pitch is poured over the top of the electrolyte. This seals the cell and binds the contents solidly together. Just before the sealing compound hardens, one or two holes are made in it so that the gases may escape.

The composition of the electrolyte itself is not exactly alike in all dry cells, as the various manufacturers follow their own particular formulas. However, as you may be curious to know something about it, we would state that one formula embraces flour, water, plaster of Paris, granulated carbon, zinc chloride, ammonium chloride, and manganese binoxide.

You will remember that the Leclanché and the dry batteries are purely open-circuit cells, and that they can be used to advantage for electric bells, annunciators, burglar alarms, gas ignition, etc., where *the current of electricity is not doing* continuous work, but only for a few seconds at a time. Consequently, the batteries have a little rest in between, if only for a few seconds.

Now, if we were to attempt to use open-circuit batteries for electric lights or motors, where the electricity must work constantly every second, the batteries would "polarize"—that is to say, they would only work a few minutes and then stop, because the chemicals used in them are of that kind that they will only allow the battery to do a little work at a time.

The batteries we have been describing will do the ordinary work for which they are intended for sometimes a year without requiring any attention, but if we try to make them do work for which they were not intended, they would only last a few days.

If we should want to operate electric lights or motors continuously from a battery we must, therefore, use

CLOSED-CIRCUIT BATTERIES

There is a great variety of ways in which closed-circuit batteries are made, but, as the main principles are very much alike, we will only describe two general kinds, those with and those without a porous cup.[2]

In the first place, we must state that closed-circuit batteries proper usually consist of a glass jar and two elements—carbon and zinc. Sometimes a porous cup is used; for what reason you will soon learn.

The chemicals that are used are usually different from those used in the open-circuit batteries and are much stronger. These chemicals are usually sulphuric acid and bichromate of potash (or chromic acid), which are mixed with water.

We will now examine two of the types of closed-circuit batteries, taking first the one without the porous cup, of which the Grenet is a good example.

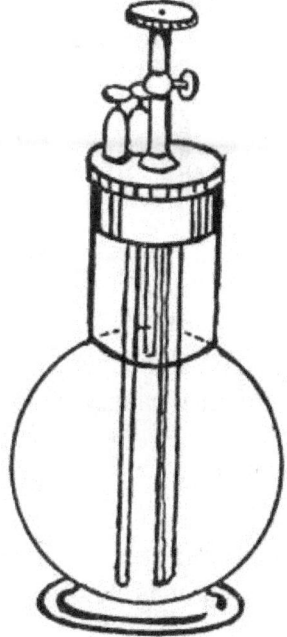

Fig. 30

This battery, as you see, consists of a glass jar, in which are placed two plates of carbon and one of zinc. (Fig. 30.) The latter is between the two carbon plates and is movable up and down, so that it may be drawn up out of the solution when it is not desired to use the battery. When the zinc is in the solution there is a steady and continuous current of electricity developed, which can be taken away by wires from the connections on top of the battery.

If the zinc were left in the solution when the battery was not in use, the acid would act upon it almost as much as though the electricity were not being used, and thus the zinc would be eaten away and the acid would be neutralized, so that no more action could be had when we wanted more electricity.

Now, in the Grenet battery we can light a lamp or run a motor for several hours continuously, but at the end of that time the solution would become black and it would do no more work. Then we must throw out that solution and put in fresh, and the battery will do the same work again, and so on.

If you should only want to light your lamp or run your motor for a few minutes, you could pull the zinc up from the solution and put it down again when you wanted the electricity once more. The carbon element in the battery is not consumed by the acid, although the zinc is.

Fig. 31

Now you will see the use of the porous cup. We will take as an illustration of this type an ordinary battery in which a porous cup is used. (Fig. 31.)

Here, you will see, the carbon is placed in the porous cup, while the zinc is outside in the glass jar. In the glass cell with the zinc is usually used water made slightly acid, and the strong solution of sulphuric acid and bichromate of potash (or chromic acid) is poured in the porous cup, where the carbon is placed.

The strong solution penetrates the porous cup very slowly and gets to the zinc, when it immediately produces a current of electricity. But the acid does not get at the zinc so freely as it does in the battery without a porous cup, and, consequently, neither the acid nor the zinc is so rapidly used up.

Where porous cups are used, the batteries will give a continuous current for a very much longer time than without them, and will, sometimes, give many hours' work every day for several months without requiring any change of solution.

Polarization.—There is one other reason why a longer working time can be had from a battery with a porous cup, and that is, in a battery without a porous cup the action of the acid upon the zinc is so rapid that the carbon plates become covered with gas, and, therefore, the proper action by the acid cannot take place upon them. Thus, the battery ceases to work, and is said to be "polarized." When a porous cup is used, the action of the acid upon the zinc is slow enough to give off only a small amount of gas, and thus the acid has a chance to act upon the carbon plates and develop a steady current of electricity.

THE WORK DONE BY BATTERIES

The pressure and quantity of electricity given off continuously by open and closed circuit batteries is very different.

The pressure (or "electromotive force") of one cell of an ordinary open-circuit battery is only about one volt, and the current is usually very much less than one ampère, except in a dry cell, which may give more.

In the closed-circuit batteries described, the electromotive force of each cell is about two volts, while the current varies from 1 to perhaps 50 ampères, according to the size of the zinc and carbon plates.

It would not matter if you made one cell as big as a barrel, nor if you put in a *dozen carbons and zincs*, the *electromotive force would not exceed the volts mentioned for each type of battery*, but the *ampère capacity would be greater* than in a smaller cell on account of the larger size of the carbon and zinc plates.

Internal Resistance.—There is one other point which affects the number of ampères which can be obtained from a closed-circuit battery, and that is whether there is a large or small internal resistance in the battery itself.

This depends upon the solution which is used and the arrangement of the plates.

If there is a high resistance in the battery itself (called "internal resistance"), the electricity must do work to overcome this resistance before it can get out of the battery to do useful work through the wires, and, consequently, the capacity in ampères is limited.

If, on the other hand, there is very little resistance in the battery, the current has very little work to flow to the wires leading from the battery, and we can get a larger quantity, or greater number of ampères.

Thus, you will see that while the closed-circuit battery is the stronger, and will do all that the open-circuit battery will do, and even more, in a short time the latter, though weaker, will do about as much work for the same amount of zinc and carbon as the former, but takes a much longer time.

BATTERIES FOR ELECTRIC LIGHT

As we have explained to you, closed-circuit batteries are used for producing incandescent electric lights in small numbers, as well as for running motors.

To operate incandescent lights, a number of batteries connected together are used. The number used depends upon the pressure which the lamps require to make them give the required light. We will now explain how the batteries are connected together for this purpose.

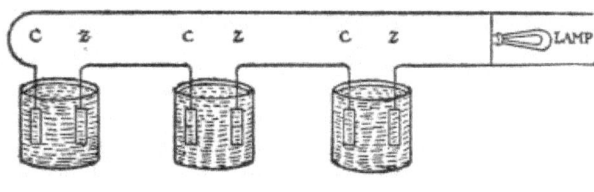

Fig. 32

Suppose you wished to light an incandescent lamp of, say, three candle-power, which required six volts. We would take three closed-circuit batteries which would each give two volts, and connect by a piece of wire the zinc of the first to the carbon of the second, and the zinc of the second to the carbon of the third, as shown in the sketch. (Fig. 32.)

We would then attach a wire to the carbon of the first and one to the zinc of the third, and there would be six volts in these two wires, which would light up one six-volt lamp nicely.

This is called connecting in series, or for intensity.

Now if each of these cells gave ten ampères alone, the three will only give ten ampères together when they are connected in series.

If our lamp only required one ampère, you would naturally think that ten similar lamps put on the wires would give as good light as the one, but that is not so.

Although you might light up two lamps, the pressure would drop and the lights would become less brilliant if you put on the whole number. So, if we wished to put on the whole ten lights we would connect another battery and thus increase the pressure, which would probably make these ten lamps burn brightly.

These rules hold good for connecting any number of batteries for lamps of any number of volts—that is to say, there should be calculated about two volts for each cell and an allowance made for drop in pressure.

CONNECTING IN MULTIPLE

There is another way of connecting batteries, and that is to obtain a larger number of ampères. This is called connecting in multiple arc, or for quantity.

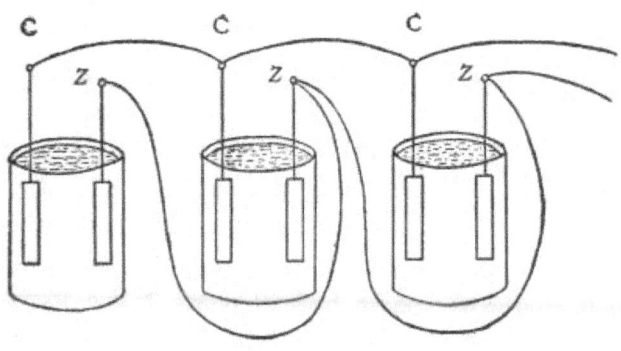

Fig. 33

Let us take again for an illustration the three cells giving each 2 volts and 10 ampères. This time we connect the carbon of the first to the carbon of the second, and the carbon of the second to that of the third; then we connect the zinc of the first to that of the second, and the zinc of the second to that of the third, as shown in the sketch. (Fig. 33.)

We then attach a wire to the zinc and one to the carbon in the third cell, and we then can obtain from these two wires *only 2 volts*, but 30 ampères.

There are, again, many ways of connecting several of these sets together, but it is not intended in this book to go into these at length, for the reason that we only set out to give a simple explanation of the first principles of this subject.

We shall therefore only give an illustration of one more method of connecting batteries which will be easy to understand. This is called

MULTIPLE SERIES

The sketch we have last given shows three batteries connected in multiple. These we will call set No. 1.

Now, suppose we take three more batteries exactly similar and connect them together just in the same manner. Let us call this set No. 2. Now take the wire leading from the carbon of set No. 2 and connect it with the wire leading from the zinc of set No. 1. Then take a wire leading from the zinc of set No. 2, and a wire leading from the carbon of set No. 1, and connect them with the lamps or motors. These two sets being connected in multiple series, we shall get 4 volts and 30 ampères.

This is called connecting in multiple series, and may be extended indefinitely with any number of batteries.

We should add that one of the elements in a battery is called "positive," and the other "negative."

THE EDISON PRIMARY BATTERY

As this type of battery will work efficiently on *either* open or closed circuit, we have thought best to describe it separately at this place, in order not to confuse your ideas while reading about batteries generally.

The type of cell we will now describe was originated by an inventor named Lalande, and was known by that name; but it has been greatly improved and rendered more efficient by Edison, and is now manufactured and sold by him under the name of the Edison Primary Battery.

Before describing the cell itself, let us consider the action that takes place in a battery of this kind.

If certain metals are placed in a suitable solution, and are connected together, outside of the solution, by wires, vigorous chemical action will take place at the surfaces of the metals, and electrical energy will be produced. The plates must be of different metals, and the solution should be one that will dissolve neither of them except when an electric current is allowed to flow.

One of the metals is usually zinc, which is gradually eaten away or dissolved by the solution while the battery is delivering electrical energy. It is the chemical combination of the zinc and the solution that produces this energy, which leaves the zinc in the form of an electric current, and passes through the solution to the other metal, out of the cell to the wire, and thence back by another wire to the zinc, where it is once more started on its circuit.

At the surface of the other metal, which may be, and frequently is, copper, small bubbles of the gas called hydrogen are produced. This gas rises to the surface of the liquid and gradually passes off into the air. But its presence offers resistance to the passage of the current; so that generally there is associated with the copper a supply of the gas oxygen. Oxygen and hydrogen are always very eager to mix with each other, and, therefore, when the hydrogen bubbles appear they are quickly taken up by the oxygen near by. The mixture of these two gases forms water, which becomes part of the solution. All of this happens so quickly that the hydrogen cannot be perceived so long as there is any oxygen left in the copper-oxide plate.

Fig. 34

In the Edison Primary Battery (Fig. 34) the plates are zinc, known as the negative, and copper oxide (copper and oxygen), or the positive. These are suspended in a solution of caustic soda and water, the plates and solution being contained in jars of glass or porcelain. The plates are provided with suitable wires for connecting the cells with one another and with the lamps, motors, or other devices which they are to operate. There are usually two zinc plates and one copper-oxide plate, or multiples thereof. The quantity of current that may be withdrawn depends on the size and number of the plates, as well as upon their construction and arrangement.

The voltage of these cells is low, being about 0.65 volt each; but this is more than compensated for by the fact that the internal resistance of the battery is so low that the voltage is not perceptibly affected even at continuous high-discharge rates, and that the voltage remains practically constant throughout the life of the cell.

Furthermore, when the battery is not in use there is practically no local action. Consequently, the cells may remain on open circuit (that is, doing no work) for years and there will be no loss of energy. The cell will then operate with the same practical efficiency as if it were new. In some classes of work this battery remains in service from four to six years without attention.

Another peculiar advantage of this battery lies in the fact that the plates and the electrolyte are so well proportioned that they are all exhausted at the same time, and then new plates and solution can be put in the jar, restoring it to its original condition. These batteries are used in great numbers for railway signal work and for other purposes, such as fire and burglar alarm systems, various telephone functions, operation of electric self-winding and programme clock systems, small electric-motor work, for low candle-power electric lamps, gas-engine ignition, electroplating, telegraph systems, chemical analysis, and other experimental work where batteries are required that will remain in use for long periods of time without requiring any attention or renewal.

The remarks that have been made on previous pages about connecting up batteries in series, multiple, and multiple series apply also to these Edison Primary Cells. Fig. 35 shows a battery of four of these cells connected in series.

SECONDARY, OR STORAGE, BATTERIES

The open and closed circuit batteries we have so far described are used to produce electricity by the action of the chemicals upon the elements contained in them. They are called primary batteries.

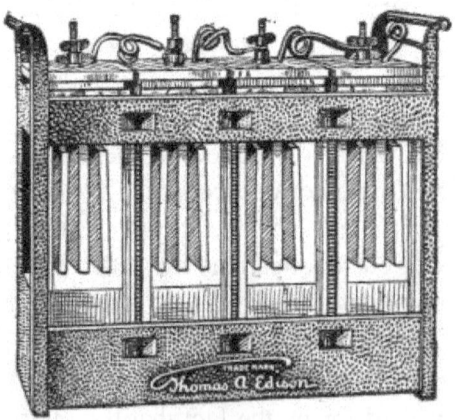

Fig. 35

The batteries which we will now tell you of are called secondary, or storage, batteries, and do not of themselves make any primary current, but simply act as reservoirs, so to speak, to hold the energy of the electric current which is led into them from a dynamo or primary battery. At the proper time and under proper conditions these secondary batteries will give back a large percentage of the energy of the electric current which has been stored in them.

This class of battery has been called by these three names: "secondary battery," "accumulator," and "storage battery"; but as the latter name is used almost exclusively in this country, we shall use it in the following description.

TWO TYPES

There are two distinct types of storage battery. One is called the "lead" or "acid" storage battery, and the other the "alkaline" or "nickel-iron" storage battery. Each of them simply acts as a reservoir to hold the energy of the electric current which is led into it, and each of them, under proper conditions, will give back that energy. As the lead storage battery is the oldest in point of discovery and invention, we will describe it first.

THE LEAD STORAGE BATTERY

A lead storage battery usually consists of a glass or hard-rubber jar containing lead plates and a solution consisting of water and sulphuric acid. A single unit is usually called a "cell." (Fig. 36.)

There are always at least two lead plates in a storage-battery cell of this kind, although there may be any number above that. For the sake of making a clearer explanation to you, we will take as an illustration a cell containing only two plates.[3]

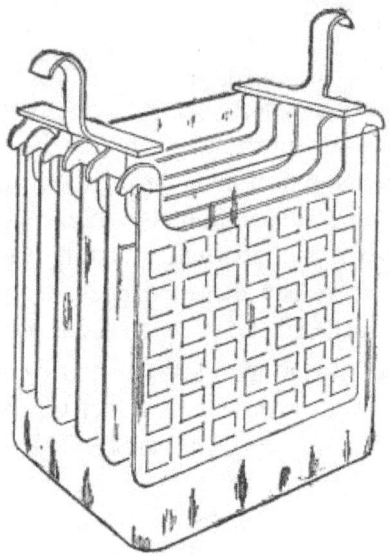

Fig. 36

We have, then, a glass or hard-rubber jar containing two lead plates and a solution consisting of water and sulphuric acid. These plates are called the "elements," and one is called the positive and the other the negative element. The solution is called the "electrolyte."

The positive element is a sheet of lead upon which is spread a paste made of red-lead. The negative element is a similar sheet of lead upon which is spread a paste made of litharge.

Now, when these plates are thus prepared, they are put into the acid solution in the jar, and a wire attached to each plate is connected with the two wires from a dynamo or other source of electric current, just as a lamp would be connected.

The electric current then goes into the storage-battery cell, entering by the positive plate and coming out by the negative. These plates and the paste upon them offer some resistance, or opposition, to the passage of the current, so the electricity must do some work to get from one to the other. The work it does in this case is to so act upon the paste that its chemical nature is changed.

So, after the primary current has been passed from one plate to the other for some time, and after several "discharges," the storage battery may be disconnected, being now "formed."

The paste on the lead plates is now found to have changed its chemical nature, the paste on the positive plate having been transformed into peroxide of lead, and that on the negative plate into spongy lead. On arriving at this condition, the paste on the plates is called "active material."

This process of "formation" is absolutely essential before the lead storage battery is ready to be used for actual work. So, when the plates have been fully "formed," the storage battery may be again connected with a source of electric current which again enters by the positive plate and leaves by the negative. This current so acts on the active material that it combines with the acid solution and, through the energy of the charging current, forms other chemical compounds which may for convenience be called "sulphates." When the charging current has flowed through the battery long enough to produce these changes in the active material the battery is said to be "charged," and is ready for useful work.

If the two wires attached to the plates are now connected with electric lamps, or a motor, or other device, the active material will develop energy in the effort to again change its nature. This energy takes the form of an electric current, which leaves the battery and passes through the conductors and operates the lamps, motors, or other devices in its passage.

In this way the battery is said to be "discharged," and at the end of its discharge it can again be charged and discharged in a similar manner for a long time, until the active material is either used up or drops off the plates.

So far as the actual details of construction are concerned, lead storage batteries are made in a great many different ways, but the materials are, in general, of the same nature as those we have mentioned above.

THE ALKALINE STORAGE BATTERY

We shall now describe an entirely different type of storage battery, which contains neither lead nor acid. It is one of the many inventions of Thomas A. Edison.

In the alkaline storage battery the gas called oxygen plays a very important part, and we will try to make it clear to you what this part is.

You are well aware of the fact that if you leave your pocket-knife out in the air it will get rusty. The reason for this is that iron or steel quickly tends to combine with the oxygen of the air, and this combination of oxygen and iron is rust, otherwise called oxide of iron, or iron oxide.

This iron oxide, or rust, is therefore the result of a chemical action between the iron and the oxygen.

Now as all chemical actions require the expenditure of energy, there has been developed either heat or electricity in the process. The oxygen may be taken away from the iron oxide, chemically; but here again would be another chemical action which would require energy to be once more expended.

Iron oxide may be made chemically in many different ways. It is frequently made in the form of a powder. Therefore, we do not have to depend upon iron rust for a supply of this material.

Before going further we must consider another oxide—namely, nickel oxide. It is characteristic of nickel that when it is combined with oxygen to a certain degree so as to form the compound known as nickel oxide, it will receive still more oxygen.

Now, if under proper conditions we compel iron oxide to give up its oxygen to some other kind of chemical compound, such as nickel oxide, we must expend energy. But, on the other hand, if this nickel oxide gives back the oxygen to the iron—which it will do if opportunity is given—there is energy produced again in receiving the oxygen. In other words, the energy previously expended, or part of it, is now returned.

This action and reaction are practically those that take place in the Edison alkaline storage battery. For simplicity of illustration we will consider a cell containing only two plates, one positive and one negative.

The negative plate is made up of a number of small, flat, perforated pockets containing iron oxide in the form of a fine powder. The positive plate is made up of small, perforated tubes containing nickel oxide mixed with very thin flakes of metallic nickel. (Fig. 37 illustrates these plates, the positive being in front.)

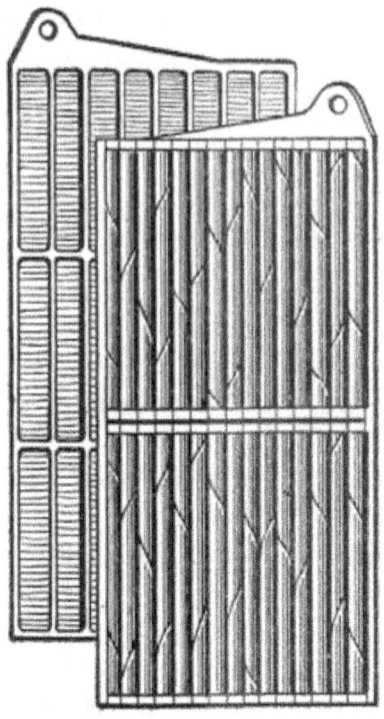

Fig. 37

These two elements, positive and negative, having wires or conductors attached, are placed in a nickeled-steel can containing the electrolyte, which consists of a potash solution. You will see that this differs from a lead storage battery, in which the electrolyte is sulphuric acid and water. If we were to put this acid solution into a metallic can (except one made of lead) the can would not last long, as the acid would quickly eat holes through it.

Now let us see what takes place in the Edison alkaline storage battery. If an electric current from a dynamo or other source of electricity is caused to pass through the positive to the negative plate the oxygen present in the iron oxide passes to and remains with the nickel oxide. During all the time this is going on the battery is said to be "charging," and when all the oxygen has been removed from the iron oxide and is taken up by the nickel oxide, then the battery is said to be "charged," and the flow of current into the battery is stopped.

A change has now taken place. The powder in the negative plate is no longer iron oxide, but has been reduced to metallic iron, because the oxygen has been removed. The powder in the positive plate is now raised to a higher or super oxide of nickel, because it has taken the oxygen that was in the iron.

But the nickel oxide will readily give up its excess of oxygen, and the iron will receive it back freely if permitted. If the proper conditions are established, this transfer of oxygen will take place, but the iron cannot receive it without delivering energy.

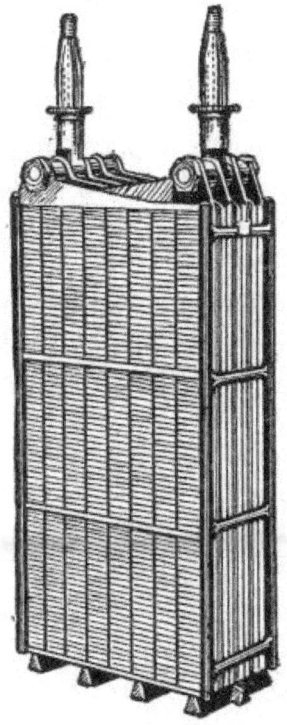

Fig. 38

The proper conditions are established by providing a conducting circuit between the two elements, in which lamps, motors, or other electrical devices are placed. As soon as this circuit is provided, the opportunity is given to the iron to receive the oxygen. This it does, and in so doing develops electrical energy.

This energy is in the form of electric current which is then delivered by the battery on what is called the "discharge," and this current may be used for lighting lamps or for operating motors or other electrical devices.

The battery is said to be discharging as long as the iron is receiving oxygen from the nickel oxide. As soon as it becomes iron oxide once more, the giving out of energy ceases and the battery is said to be "discharged," and must again be charged to obtain further work from it. Such a battery can be charged and discharged an indefinite number of times.

This type of battery is very rugged, and its combinations are not self-destructive. It is very simple, as it provides chiefly for the movement of the oxygen back and forth; besides, it gives much more current for its weight than the lead type of storage battery. (Fig. 38 shows the plates of a standard Edison cell removed from container.)

CONNECTING STORAGE BATTERIES

On the discharge, one cell of a lead storage battery gives an average of about 2 volts, and a cell of alkaline storage battery about 1.2 volts, no matter what its size or the number of plates may be. When there are more than two plates in one cell, all the positives in that cell are connected together by metallic strips or bands, and all negatives in the cell are connected together in a similar way.

Although we cannot obtain more than the above-named electromotive force from one cell of either type of storage battery, we can obtain a greater ampère capacity by using large plates instead of small ones, or by using a larger number of small size.

The same effects are produced by connecting the cells in series, or multiple, or multiple series, as we showed you in regard to primary batteries; and the storage batteries may be charged as well as discharged when connected in any one of these ways.

CHARGING CURRENT

The current which is used for charging must always be greater in pressure than that of the storage batteries which are being charged. If it is not, the storage batteries will be the stronger of the two and will overpower the charging current and so discharge themselves.

X

CONCLUSION

We will now bring this little volume to a close, having given you a brief outline of the simplest rudiments of that wonderful power of nature, Electricity.

We may compare this subject to a beautiful house the inside of which you would like to examine from top to bottom. We have opened the door for you; now walk in and examine everything. There may be a great many stairs to climb, but what you see and learn will repay for all the trouble.

THE END

www.ingramcontent.com/pod-product-compliance
Lightning Source LLC
Chambersburg PA
CBHW061203180526
45170CB00002B/937